The Structure of the Rational Concordance Group of Knots

Memoirs
of the
American Mathematical Society

Number 885

The Structure of the Rational Concordance Group of Knots

Jae Choon Cha

September 2007 • Volume 189 • Number 885 (second of 4 numbers) • ISSN 0065-9266

American Mathematical Society
Providence, Rhode Island

2000 *Mathematics Subject Classification.* Primary 57M25, 57Q45, 57Q60.

Library of Congress Cataloging-in-Publication Data

Cha, Jae Choon, 1971–
    The structure of the rational concordance group of knots / Jae Choon Cha
      p. cm. — (Memoirs of the American Mathematical Society, ISSN 0065-9266 ; no. 885)
    "September 2007, volume 189, number 885 (second of 4 numbers)."
    Includes bibliographical references.
    ISBN 978-0-8218-3993-5 (alk. paper)
    1. Low-dimensional topology. 2. Knot theory. 3. Concordances (Topology). I. Title.
QA612.14.C43 2007
514′.22—dc22   2007060801

Memoirs of the American Mathematical Society

This journal is devoted entirely to research in pure and applied mathematics.

Subscription information. The 2007 subscription begins with volume 185 and consists of six mailings, each containing one or more numbers. Subscription prices for 2007 are US$649 list, US$519 institutional member. A late charge of 10% of the subscription price will be imposed on orders received from nonmembers after January 1 of the subscription year. Subscribers outside the United States and India must pay a postage surcharge of US$38; subscribers in India must pay a postage surcharge of US$43. Expedited delivery to destinations in North America US$53; elsewhere US$130. Each number may be ordered separately; *please specify number* when ordering an individual number. For prices and titles of recently released numbers, see the New Publications sections of the *Notices of the American Mathematical Society*.

Back number information. For back issues see the *AMS Catalog of Publications*.

Subscriptions and orders should be addressed to the American Mathematical Society, P. O. Box 845904, Boston, MA 02284-5904, USA. *All orders must be accompanied by payment.* Other correspondence should be addressed to 201 Charles Street, Providence, RI 02904-2294, USA.

Copying and reprinting. Individual readers of this publication, and nonprofit libraries acting for them, are permitted to make fair use of the material, such as to copy a chapter for use in teaching or research. Permission is granted to quote brief passages from this publication in reviews, provided the customary acknowledgment of the source is given.

Republication, systematic copying, or multiple reproduction of any material in this publication is permitted only under license from the American Mathematical Society. Requests for such permission should be addressed to the Acquisitions Department, American Mathematical Society, 201 Charles Street, Providence, Rhode Island 02904-2294, USA. Requests can also be made by e-mail to reprint-permission@ams.org.

Memoirs of the American Mathematical Society is published bimonthly (each volume consisting usually of more than one number) by the American Mathematical Society at 201 Charles Street, Providence, RI 02904-2294, USA. Periodicals postage paid at Providence, RI. Postmaster: Send address changes to Memoirs, American Mathematical Society, 201 Charles Street, Providence, RI 02904-2294, USA.

© 2007 by the American Mathematical Society. All rights reserved.
Copyright of this publication reverts to the public domain 28 years
after publication. Contact the AMS for copyright status.
This publication is indexed in *Science Citation Index*®, *SciSearch*®, *Research Alert*®,
CompuMath Citation Index®, *Current Contents*®/*Physical, Chemical & Earth Sciences*.
Printed in the United States of America.

∞ The paper used in this book is acid-free and falls within the guidelines
established to ensure permanence and durability.
Visit the AMS home page at http://www.ams.org/

10 9 8 7 6 5 4 3 2 1   12 11 10 09 08 07

Dedicated to the memory of
Jerome P. Levine
(May 4, 1937–April 8, 2006)

Contents

Acknowledgments   ix

Chapter 1.   Introduction   1
   1.1.   Integral and rational knot concordance   2
   1.2.   Main results   4

Chapter 2.   Rational knots and Seifert matrices   11
   2.1.   Generalized Seifert surfaces   11
   2.2.   Limits of Seifert matrices   16

Chapter 3.   Algebraic structure of $\mathcal{G}_n$ 21
   3.1.   Invariants of Seifert matrices   21
   3.2.   Invariants of limits of Seifert matrices   25
   3.3.   Computation of $e(\mathcal{A})$ 31
   3.4.   Artin reciprocity and norm residue symbols   36
   3.5.   Computation of $d(\mathcal{A})$ 39

Chapter 4.   Geometric structure of $\mathcal{C}_n$ 47
   4.1.   Realization of rational Seifert matrices   47
   4.2.   Construction of slice disks in rational balls   51
   4.3.   Rational and integral concordance   59
   4.4.   Subrings of rationals   65

Chapter 5.   Rational knots in dimension three   67
   5.1.   Rational (0)- and (0.5)-solvability   67
   5.2.   Effect of complexity change   76
   5.3.   Realization of Alexander modules by ribbon knots   82
   5.4.   Knots which are not rationally (1.5)-solvable   86

Bibliography   93

Abstract

We study the group of rational concordance classes of codimension two knots in rational homology spheres. We give a full calculation of its algebraic theory by developing a complete set of new invariants. For computation, we relate these invariants with limiting behaviour of the Artin reciprocity over an infinite tower of number fields and analyze it using tools from algebraic number theory. In higher dimensions it classifies the rational concordance group of knots whose ambient space satisfies a certain cobordism theoretic condition. In particular, we construct infinitely many torsion elements. We show that the structure of the rational concordance group is much more complicated than the integral concordance group from a topological viewpoint. We also investigate the structure peculiar to knots in rational homology 3-spheres. To obtain further nontrivial obstructions in this dimension, we develop a technique of controlling a certain limit of the von Neumann L^2-signature invariants.

2000 *Mathematics Subject Classification.* Primary 57M25, 57Q45, 57Q60.
Key words and phrases. Knots, Rational Homology Spheres, Concordance.

Acknowledgments

I would like to express my appreciation of help received from several people in pursuing this project. I would particularly like to thank Kent Orr for numerous valuable conversions about this work. Discussions with Tim Cochran, Jim Davis, Ki Hyoung Ko, Michael Larsen, and Jerome Levine were also very interesting and helpful.

CHAPTER 1

Introduction

In this paper we study a classification problem of knots in rational homology spheres. More precisely, a closed manifold with the rational homology of the sphere of the same dimension is called a *rational sphere*, and a codimension two locally flat sphere embedded in a rational sphere is called a *rational knot*. Two rational knots K and K' with the same dimension are said to be *(rationally) concordant* if there is a rational homology cobordism between their ambient spaces which contains a locally flat annulus bounded by $K \cup -K'$. Under connected sum, concordance classes of n-dimensional rational knots form an abelian group which we call the *rational knot concordance group* and denote by $\mathcal{C}_n$. Our main aim is to study the structure of $\mathcal{C}_n$.

There are some interesting motivations of our research. We list a few of them below. First, rational knot concordance has a close relationship with concordance of links in the ordinary sphere. While there is a known framework of the study of link concordance (e.g., see Cappell–Shaneson [3] and Le Dimet [30]), it still remains far more elusive than knot concordance because of a lack of our understanding of related homotopy theoretical and surgery theoretical problems. In the remarkable work of Cochran and Orr [10, 12], it was first proposed that problems on link concordance can be transformed into ones on rational knot concordance. Based on this idea they proved the long-standing conjecture that not all links are concordant to boundary links. For some further developments and applications, see subsequent work of Ko and the author [7, 8].

Since these successful applications, more systematic study of rational concordance has been called on. Note that rational knot concordance is a natural generalization of ordinary concordance of knots in the sphere which has a deep and rich theory. For results on ordinary knot concordance particularly related to this paper, see Levine [33, 32], Kervaire [26], Cappell–Shaneson [3], Casson–Gordon [5, 6], and Cochran–Orr–Teichner [13, 14]. Regarding this, it is natural to ask whether one can establish an analogous theory for rational knots. The only result which has been known is that the knot signature function [45, 38, 44, 37, 33] extends to rational knots [12, 7]. Indeed all the known applications to link concordance depend on this signature invariant.

Received by editor November 5, 2004

We also remark that Cochran and Orr pointed out in an unpublished note that rational concordance is closely related with the rational homology surgery theory developed by Quinn [**39**] and Taylor and Williams [**43**]. From this viewpoint, rational concordance can be regarded as a particular instance of rational homology surgery which provides computational techniques and examples.

In this paper we perform through analysis of the structure of the rational knot concordance group $\mathcal{C}_n$. As one of the results, we give a full calculation of its algebraic theory by constructing new algebraic invariants and computing them. In particular, we discover infinitely many independent finite order elements in $\mathcal{C}_n$ for odd $n > 1$. Compared with the ordinary knot concordance group from a topological viewpoint, it turns out that the rational knot concordance group has a very different structure. We also investigate the structure peculiar to knots in rational 3-spheres. In this dimension we develop a computational technique of further obstructions to rational concordance using the von Neumann ρ-invariants.

In this paper we work in the category of oriented piecewise linear manifolds. Submanifolds are always assumed to be locally flat. Our results also hold in the categories of smooth and topological manifolds with minor modifications if necessary.

1.1. Integral and rational knot concordance

We begin by recalling known results on the group of concordance classes of codimension two knots in S^{n+2}, which we call the *integral* knot concordance group and denote by $\mathcal{C}_n^{\mathbb{Z}}$. In higher dimensions, $\mathcal{C}_n^{\mathbb{Z}}$ is classified using abelian invariants of knots which can be extracted using several different techniques. For $n > 1$, Kervaire [**26**] and Levine [**33, 32**] first computed the structure of $\mathcal{C}_n^{\mathbb{Z}}$ using Seifert surfaces and Seifert matrices. Cappell and Shaneson applied their homology surgery theory to identify $\mathcal{C}_n^{\mathbb{Z}}$ with a surgery obstruction Γ-group [**3**]. In [**23, 24**], Kearton showed that the same classification can be obtained using the Blanchfield form [**1**] for odd $n > 1$. We remark that $\mathcal{C}_n^{\mathbb{Z}}$ is isomorphic to the set of integral homology concordance classes of knots in integral homology spheres for $n > 1$. This justifies our terminology "integral (homology) concordance".

By work of Cochran, Orr, Ko, and the author [**12, 7, 8**], a framework of the rational concordance theory has been initiated. Its basic strategy is similar to the above integral theory, but, it involves more sophisticated topological and algebraic constructions that produce objects whose structures have not been fully understood.

The essential difference between the integral and rational theories is illustrated in the following general observation which introduces the notion of *complexity*. Suppose that M is a properly embedded connected submanifold of codimension two in W. If $H_1(W;\mathbb{Z}) = H_2(W;\mathbb{Z}) = 0$, then from the Alexander duality, it follows that $H_1(W - M;\mathbb{Z})$ is the infinite cyclic group $\mathbb{Z}$ generated by a meridian of M. On the other hand, if we assume a weaker condition that $H_1(W;\mathbb{Q}) = H_2(W;\mathbb{Q}) = 0$, instead of the integral homology condition, then the torsion-free part of $H_1(W - M;\mathbb{Z})$ is still $\mathbb{Z}$, but in general, the meridian no longer generates it. The duality (with rational coefficients) merely says that the meridian represents a nonzero element in $\mathbb{Z}$, and its absolute value c measures the extent of this failure. We call c the *complexity*. Because the complexity can be an arbitrarily large integer, one cannot apply key arguments of the integral concordance theory to the rational case. For example, suppose that an n-dimensional knot K in S^{n+2} is a slice knot, that is, there exists a pair (Δ, D) of an (integral homology) $(n+3)$-ball Δ and an embedded $(n+1)$-disk D such that $\partial(\Delta, D) = (S^{n+2}, K)$. From the above observation it follows that the abelianization homomorphism $\pi_1(S^{n+2} - K) \to \mathbb{Z}$ extends to $\pi_1(\Delta - D^2)$. This enables us to extract obstructions to being an integral slice knot from abelian invariants of knots. In contrast to this, if Δ were a rational ball, then the homomorphism would not extend in general. In this case it can be seen that a homomorphism $\pi_1(S^{n+2} - K) \to \mathbb{Z}$ sending a meridian to the complexity c of $D \subset \Delta$ does extend. But the value of c is unknown unless a particular (Δ, D) is given.

From the viewpoint of [**7**] based on Seifert surfaces and Seifert matrices, the notion of complexity discussed above is also related to the fact that a rational knot may not admit any Seifert surface while every integral knot does. Instead, for a rational n-dimensional knot K with $n > 1$, there is a positive integer c such that the union of c parallel copies of K bounds an embedded submanifold, which is called a *generalized Seifert surface of complexity c* (a framing condition is required for $n = 1$; see Section 2.1 for a precise description).

If n is odd, a Seifert matrix of a generalized Seifert surface is defined in the usual way. A difference from the integral theory is that our Seifert matrix has rational entries in general. Using Seifert matrices, we can define an algebraic analogue of $\mathcal{C}_n$, which is called the *algebraic rational concordance group* and denoted by $\mathcal{G}_n$, together with a homomorphism $\Phi_n \colon \mathcal{C}_n \to \mathcal{G}_n$. This may also be viewed as an analogue of Levine's homomorphism of $\mathcal{C}_n^{\mathbb{Z}}$ into the *algebraic (integral) concordance group* of Seifert matrices. Roughly speaking, for each c, we form the algebraic concordance group $G_{n,c}$ of Seifert matrices of generalized Seifert surfaces of complexity c as Levine did in [**33**], and define $\mathcal{G}_n = \varinjlim_c G_{n,c}$ to be the limit of a direct system consisting of the $G_{n,c}$ and certain homomorphisms. This construction can be viewed as a functorial image of $\mathbb{Q} = \varinjlim_c (1/c)\mathbb{Z}$. (In Section 2.2, a purely algebraic definition of $\mathcal{G}_n$ is given.)

Given a rational n-dimensional knot, a Seifert matrix of a generalized Seifert surface of complexity c represents an element in $G_{n,c}$, and sending it by the canonical homomorphism into the limit, an element in $\mathcal{G}_n$ is obtained. For odd $n > 1$, it gives rise to the homomorphism $\Phi_n \colon \mathcal{C}_n \to \mathcal{G}_n$. For $n = 1$, it turns out that a homomorphism Φ_1 into $\mathcal{G}_1$ is defined on a subgroup $s\mathcal{C}_1$

of $\mathcal{C}_1$ which fits into an exact sequence
$$0 \longrightarrow s\mathcal{C}_1 \longrightarrow \mathcal{C}_1 \longrightarrow \mathbb{Q}/\mathbb{Z} \longrightarrow 0.$$
(For details, see Section 2.1.) In [12, 7] it was shown that $\mathcal{C}_n$ contains a subgroup isomorphic to $\mathbb{Z}^\infty$ by investigating a signature invariant of $\mathcal{G}_n$ and pulling it back via Φ_n.

In spite of the importance of $\mathcal{G}_n$ and Φ_n in the study of $\mathcal{C}_n$, several interesting questions on their structures have not been answered. For example, it has not been known whether $\mathcal{G}_n$ and $\mathcal{C}_n$ have torsion elements. Also there has been no geometric answer to the question how much structure of $\mathcal{C}_n$ can be revealed via Φ_n.

1.2. Main results

1.2.1. The structure of $\mathcal{G}_n$.
As an answer to the above questions, we give a full calculation of the structure of the limit $\mathcal{G}_n$.

THEOREM 1.1. *The group $\mathcal{G}_n$ is isomorphic to $\mathbb{Z}^\infty \oplus (\mathbb{Z}/2)^\infty \oplus (\mathbb{Z}/4)^\infty$.*

Although it is abstractly isomorphic to the integral (algebraic) knot concordance group of Levine, we do not have any natural identification which is topologically meaningful. In fact it turns out that, from a topological viewpoint, their structures are drastically different. It will be discussed in a later subsection.

In Chapter 3, we construct a complete set of invariants of $\mathcal{G}_n$, and by realizing and computing them, we prove Theorem 1.1. Briefly, our invariants of $\mathcal{G}_n$ can be described as follows. We need to start with known invariants of the integral algebraic concordance group $G_{n,c}$. An algebraic number z is called *reciprocal* if z and z^{-1} are conjugate, i.e., if they share the same irreducible polynomial over $\mathbb{Q}$. It is known that the concordance group of Seifert matrices maps into the direct sum of Witt groups of nonsingular hermitian forms on finite dimensional vector spaces over number fields $\mathbb{Q}(z)$ equipped with the involution $\bar{z} = z^{-1}$, where z runs over reciprocal numbers. This associates to a Seifert matrix A a Witt class of a hermitian form A_z over $\mathbb{Q}(z)$, which is called the z-primary part of A. The signature of A_z (defined for $|z| = 1$ only), the modulo 2 residue class of the dimension r of A_z, and the discriminant

$$\text{dis } A_z = (-1)^{\frac{r(r+1)}{2}} \det A_z \in \frac{\mathbb{Q}(z+z^{-1})^\times}{\{u\bar{u} \mid u \in \mathbb{Q}(z)^\times\}}$$

give rise to invariants of the integral algebraic concordance group.

To construct invariants of $\mathcal{G}_n$, we take "limits" of the above invariants. Let P be the set of all sequences $\alpha = (\ldots, \alpha_2, \alpha_1)$ of reciprocal numbers α_i such that $(\alpha_{ri})^r = \alpha_i$ for all i and r. (P can be viewed as the limit of an inverse system consisting of the sets of reciprocal numbers and morphisms $z \to z^r$.) Let P_0 be its subset consisting of $\alpha = (\alpha_i)$ with $|\alpha_i| = 1$. For an element $\mathcal{A}$ in $\mathcal{G}_n$ represented by a Seifert matrix A of complexity c, we consider the invariants of A associated to the c-th coordinate α_c of $\alpha \in P$ (or P_0). That is, we define

$$s(\mathcal{A}) = (\text{signature of } A_{\alpha_c})_{\alpha \in P_0} \qquad \in \mathbb{Z}^{P_0},$$
$$e(\mathcal{A}) = (\text{dimension of } A_{\alpha_c} \bmod 2)_{\alpha \in P} \in (\mathbb{Z}/2)^P.$$

A discriminant invariant of $\mathcal{G}_n$ is also defined in a similar way, but its value lives in a more complicated object since the codomain of the discriminant of A_z depends on z. We form a limit

$$\varinjlim_{i} \prod_{\alpha \in P} \frac{\mathbb{Q}(\alpha_i + \alpha_i^{-1})^\times}{\{u\bar{u} \mid u \in \mathbb{Q}(\alpha_i)^\times\}},$$

and define the third invariant $d(\mathcal{A})$ to be the element in the limit represented by

$$(\text{dis } A_{\alpha_c})_{\alpha \in P} \in \prod_{\alpha \in P} \frac{\mathbb{Q}(\alpha_c + \alpha_c^{-1})^\times}{\{u\bar{u} \mid u \in \mathbb{Q}(\alpha_c)^\times\}}.$$

We remark that the above invariants of $\mathcal{A}$ do not carry full information on (the concordance class of) the representative A. Indeed, observing that α_c has the property that for any r there exists a reciprocal r-th root $(=\alpha_{rc})$, it can be seen that not all reciprocal numbers appear as the concerned parameter α_c. An interesting result is that this limited information gives rise to well-defined invariants of $\mathcal{G}_n$, and furthermore, it is enough to classify $\mathcal{G}_n$.

THEOREM 1.2. *The invariants s, e, and d form a complete set of invariants of $\mathcal{G}_n$.*

Pulling back via $\Phi_n \colon \mathcal{C}_n \to \mathcal{G}_n$, s, e, and d give rise to invariants of the rational knot concordance group.

In Section 3.2 we discuss the above construction in detail and prove Theorem 1.2. We remark that our invariant $s(\mathcal{A})$ is equivalent to the signature invariants studied in [12, 7]. Compared with other invariants e and d, the signature s is much easier to define and use, since the crucial condition of $\alpha = (\alpha_i)$ that α_i must have a reciprocal r-th root for all r is automatically satisfied whenever α_i has unit length.

In the proof of Theorem 1.1, concrete examples of infinitely many independent order 2 and 4 elements in $\mathcal{G}_n$ are constructed. Some order 2 elements can be detected by using the invariant e. Since its value lives in a simple domain $(\mathbb{Z}/2)^P$, it is easier to handle than d. The crucial step is to find elements α in P which are not contained in P_0 so that the α_c-primary parts have no contribution to the signature. (See Corollary 3.19.)

For order 4 elements, much more complicated algebraic arguments are involved because we must compute the invariant d that lives in a limit. The Artin reciprocity, which is one of the central machinery in algebraic number theory, plays a crucial role in our computation. In what follows we discuss our idea briefly. In order to show the nontriviality of d in its codomain, we need to investigate the norms of field extensions of the form

$\mathbb{Q}(\alpha_c)/\mathbb{Q}(\alpha_c+\alpha_c^{-1})$ where $\alpha \in P$, and study their limiting behaviour as c goes to infinity. For a fixed c, by the Hasse principle, this global problem is reduced into a local problem over completions $\mathbb{Q}(\alpha_c)^v/\mathbb{Q}(\alpha_c+\alpha_c^{-1})_v$ with respect to valuations v of $\mathbb{Q}(\alpha_c+\alpha_c^{-1})$. Now we appeal to the local Artin reciprocity, which asserts that there is an epimorphism

$$\mathbb{Q}(\alpha_c+\alpha_c^{-1})_v^\times \longrightarrow \mathrm{Gal}(\mathbb{Q}(\alpha_c)^v/\mathbb{Q}(\alpha_c+\alpha_c^{-1})_v)$$

whose kernel consists of nonzero norms of $\mathbb{Q}(\alpha_c)^v/\mathbb{Q}(\alpha_c+\alpha_c^{-1})_v$. We investigate the limiting behaviour of the effect of this local Artin map on the discriminant, as c goes to infinity. In general, it seems a hard algebraic problem requiring a deep understanding of number theoretic phenomena. Fortunately, by constructing Seifert matrices carefully, we are led to a very specific instance of the problem for which we can control the limiting behaviour. One of the key steps is to find suitable valuations v. For this we do inductive analysis of prime splitting over a tower of field extensions of an arbitrary height (see Section 3.5 for details). This enables us to construct desired finite order elements in $\mathcal{G}_n$.

1.2.2. The structure of Φ_n. Sections 4.1 and 4.2 of this paper are devoted to a geometric study of the structure of the homomorphism Φ_n. An obvious observation is that Φ_n is not injective; for instance, Φ_n does not detect the effect of the action of the rational homology cobordism group of rational $(n+2)$-spheres on $\mathcal{C}_n$ given by connected sum with ambient spaces. In order to avoid such complications from ambient spaces, we think of a subgroup $b\mathcal{C}_n$ of $\mathcal{C}_n$ which is generated by knots in rational spheres bounding parallelizable rational balls. The following result shows that this geometrically defined subgroup $b\mathcal{C}_n$ is crucial in understanding the homomorphism Φ_n; in higher odd dimensions, it is a largest subgroup which is classified by Φ_n.

THEOREM 1.3.
 (1) *For even n, $b\mathcal{C}_n$ is trivial.*
 (2) *For odd $n > 3$, the restriction $\Phi_n|_{b\mathcal{C}_n} : b\mathcal{C}_n \to \mathcal{G}_n$ is an isomorphism.*
 (3) *For $n = 3$, $\Phi_3|_{b\mathcal{C}_3}$ is an isomorphism of $b\mathcal{C}_3$ onto an index two subgroup of $\mathcal{G}_3$ which is isomorphic to $\mathbb{Z}^\infty \oplus (\mathbb{Z}/2)^\infty \oplus (\mathbb{Z}/4)^\infty$.*
 (4) *For $n = 1$, $\Phi_1|_{b\mathcal{C}_1 \cap s\mathcal{C}_1}$ is a surjection onto $\mathcal{G}_1$.*

This can be compared with the results of Kervaire [26] and Levine [33] on integral knot concordance. In the topological category, the above (2) holds for $n = 3$ instead of (3). An immediate consequence of Theorem 1.3 is that for odd $n > 1$ the exact sequence

$$0 \longrightarrow \mathrm{Ker}\,\Phi_n \longrightarrow \mathcal{C}_n \xrightarrow{\Phi_n} \mathrm{Im}\,\Phi_n \longrightarrow 0$$

splits, and therefore, $b\mathcal{C}_n \cong \mathrm{Im}\,\Phi_n \cong \mathcal{G}_n$ (or its index two subgroup if $n=3$) is a direct summand of $\mathcal{C}_n$. From this it follows that $\mathcal{C}_n$ contains infinitely many independent elements of order 2, 4, and infinite.

In Section 4.1, the surjectivity is shown by a constructive realization algorithm of Seifert matrices of rational knots. In general, our algorithm produces a knot in an ambient space which is not necessarily an integral homology sphere, since a given Seifert matrix may have non-integral entries. We also remark that a characterization theorem of Alexander polynomials of rational Seifert matrices is proved in Section 3.5 (see Theorem 3.22).

Section 4.2 is devoted to the injectivity. For this we mainly use techniques of ambient surgery. In comparison with the integral case, we need more complicated arguments because we should perform ambient surgery in a rational ball which may have nontrivial homotopy groups even below the middle dimension.

We remark that when n is even our argument shows more than Theorem 1.3 (1). For more details, see Theorem 4.13.

1.2.3. Comparison with the integral knot concordance group. As remarked above, although $b\mathcal{C}_n \cong \mathcal{C}_n^{\mathbb{Z}}$ for $n > 1$, we have no canonical identification between them. A natural topological way to compare their structures is to study the canonical map $\mathcal{C}_n^{\mathbb{Z}} \to b\mathcal{C}_n \subset \mathcal{C}_n$. In Section 4.3, using our invariants of $\mathcal{G}_n$, we prove the following results:

THEOREM 1.4.
  (1) For odd n, the kernel of $\mathcal{C}_n^{\mathbb{Z}} \to b\mathcal{C}_n$ contains a subgroup isomorphic to $(\mathbb{Z}/2)^{\infty}$.
  (2) For odd $n > 1$, the cokernel of $\mathcal{C}_n^{\mathbb{Z}} \to b\mathcal{C}_n$ contains a summand isomorphic to $\mathbb{Z}^{\infty} \oplus (\mathbb{Z}/2)^{\infty} \oplus (\mathbb{Z}/4)^{\infty}$.

This illustrates that the geometric structures of integral and rational knot concordance groups are drastically different. Cochran (using work of Fintushel and Stern [17]) showed that the figure eight knot is a nontrivial element in the kernel of $\mathcal{C}_1^{\mathbb{Z}} \to b\mathcal{C}_1$, and Kawauchi showed that an analogous higher dimensional knot is nontrivial and contained in the kernel of $\mathcal{C}_n^{\mathbb{Z}} \to b\mathcal{C}_n$ for $n = 4k + 1 > 1$ [22]. Theorem 1.4 (1) is a generalization of these results. Theorem 1.4 (2) is a generalization of a previous result of Ko and the author [7].

Most of our results in higher dimensions extend to the case of knots in R-homology spheres for any subring R of $\mathbb{Q}$. This is discussed in Section 4.4.

In an unpublished note by Cochran and Orr, they studied rational concordance in higher dimensions using the rational homology surgery theory due to Quinn [39] and Taylor and Williams [43]. Our algebraic results can be viewed as computation of surgery obstruction Γ-groups which are related to their work. (See Remarks 3.42 and 4.20.)

1.2.4. Knots in rational 3-spheres. From the above results, it follows that the concordance class of a knot representing an element in $b\mathcal{C}_n$ is determined by its Seifert matrix for odd $n > 1$. However, it is not true for $n = 1$:

THEOREM 1.5. *There exist knots in S^3 which have Seifert matrices of integral slice knots but are not rational slice knots.*

This can be viewed as a generalization of the result of Casson and Gordon [**6, 5**] that the integral concordance classes of knots in S^3 are not determined by Seifert matrices. More recently, Cochran, Orr, and Teichner [**13, 14**] have developed a new obstruction to being an integral slice knot. Although they partially considered rational concordance, no information on the structure of $\mathcal{C}_1$ was extracted via their obstruction because of the same sort of difficulty that the complexity may be nontrivial.

In Chapter 5, we extend methods and results of Cochran–Orr–Teichner [**13, 14**] to rational concordance. These results hold in the topological category (where submanifolds are assumed to be locally flat) as well as the piecewise linear and smooth categories. To discuss out results, we recall some basic ideas of the work of Cochran–Orr–Teichner. For $h = 0, 0.5, 1, 1.5, \ldots$, they define a *rational (h)-solution* of a closed 3-manifold M to be a 4-manifold W bounded by M whose intersection form over a solvable group ring coefficient satisfies certain surgery theoretic conditions. When a rational knot K admits a generalized Seifert surface, it determines a well-defined zero-framing, and we can think of the zero-surgery manifold M of K. If M admits a rational (h)-solution W, we say K is *rationally (h)-solvable*. This is a refinement of the rational slice condition; a rational solution W can be viewed as an "approximation" of a slice disk exterior in a rational 4-ball. Concordance classes of rationally (h)-solvable knots form a subgroup $\mathcal{F}_{(h)}^{\mathbb{Q}}$ of $\mathcal{C}_1$ which gives a filtration

$$\{0\} \subset \cdots \subset \mathcal{F}_{(n.5)}^{\mathbb{Q}} \subset \mathcal{F}_{(n)}^{\mathbb{Q}} \subset \cdots \subset \mathcal{F}_{(0.5)}^{\mathbb{Q}} \subset \mathcal{F}_{(0)}^{\mathbb{Q}} \subset s\mathcal{C}_1 \subset \mathcal{C}_1.$$

(In [**13**], rational solvability was considered only for knots in S^3. In Section 5.1, we give a reformed version of the original definition in [**13**] for rational knots.)

We investigate the structure of this filtration. First, in Section 5.1, we give a characterization of rationally (0)- and (0.5)-solvable knots. For $h = 0$, it turns out that the rational (h)-solvability of a knot is indeed none more than a condition on its ambient space; a knot is rationally (0)-solvable if and only if its ambient space admits a rational (0)-solution. In general, the solvability condition of the ambient space is not sufficient, since there are further complications from knotting. For $h = 0.5$, it turns out that the limit of Seifert matrices gives rise to an obstruction to being a rationally (0.5)-solvable knot. In fact in Section 5.1 we prove the following result:

THEOREM 1.6. $\mathcal{F}_{(0)}^{\mathbb{Q}}/\mathcal{F}_{(0.5)}^{\mathbb{Q}} \cong \mathcal{G}_1 \cong \mathbb{Z}^\infty \oplus (\mathbb{Z}/2)^\infty \oplus (\mathbb{Z}/4)^\infty.$

For a further investigation of the structure of the filtration, we use the von Neumann ρ-invariants of Cheeger–Gromov [**9**] which were first considered by Cochran–Orr–Teichner [**13, 14**] for knots. This enables us to prove the following result:

THEOREM 1.7. $\mathcal{F}_{(1)}^{\mathbb{Q}}/\mathcal{F}_{(1.5)}^{\mathbb{Q}}$ *has infinite rank.*

In fact we construct integral knots in S^3 with metabolic Seifert matrices which generate an infinite rank subgroup in $\mathcal{F}_{(1)}^{\mathbb{Q}}/\mathcal{F}_{(1.5)}^{\mathbb{Q}}$ (see Theorem 5.25). From this Theorem 1.5 follows.

In [**13, 14**], Cochran–Orr–Teichner showed that certain von Neumann ρ-invariants are obstructions to having a rational $(n.5)$-solution of a *given* complexity c (n is an integer), where the complexity of a rational solution W is defined in a similar way as the general discussion in Section 1.1. The essential problem in applying Cochran–Orr–Teichner's idea to rational concordance is that the obstruction depends on the value of c which can be an arbitrary positive integer. Their main results on integral concordance in [**13, 14**] are obtained by considering the special case of $c = 1$.

The main idea of the proof of Theorem 1.7 is to control the concerned von Neumann ρ-invariant as c varies. Very roughly speaking, the von Neumann ρ-invariants are determined by elements in a metabolizer of the Blanchfield form on a certain Alexander module, where the structures of the Alexander module, Blanchfield form, and the metabolizer configuration depend on the value of c. To handle an arbitrary value of c, we investigate the limiting behaviour of them as c goes to infinity. Even in the case of integral knot concordance, calculation of the configuration of metabolizers is the most difficult step in applying this machinery. Our contribution in this regard is to give a concrete construction of knots with the following property: *for any c there is a unique proper nontrivial submodule in the Alexander module associated to complexity c.* This enables us to compute explicitly the metabolizer for any complexity c and to control the behaviour of the von Neumann ρ-invariant. In fact, we show that there is a family of infinitely many knots such that a particular von Neumann ρ-invariant, which is independent of c, gives an obstruction to admitting a rational solution of any complexity c (see Theorem 5.21).

CHAPTER 2

Rational knots and Seifert matrices

2.1. Generalized Seifert surfaces

In this section we discuss a generalization of Seifert surfaces to the rational knot case. We will sometimes contrast the sophistication peculiar to rational knots by comparing it with the counterpart in integral concordance theory. Basically most of the ideas of this section are from [7]. We supplement some technical reformulations for use in later sections. While [7] deals with a general theory of rational *links*, we focus on the case of knots. We remark that in the work of Cochran and Orr [12] Blanchfield forms were used instead of Seifert matrices. Our approach will be particularly useful for the geometric study of rational knots, as well as for practical computation of algebraic invariants. For instance, we will give a constructive realization algorithm of rational Seifert matrices in Section 4.1.

DEFINITION 2.1. An embedded n-sphere in a rational $(n+2)$-sphere is called a *(rational) n-knot*.

Recall that we always assume that all submanifolds are locally flat. Sometimes we view a rational knot K in Σ as a manifold pair (Σ, K). When the ambient space of a knot is the honest sphere S^{n+2}, we occasionally call it an *integral knot*.

DEFINITION 2.2 ([7]). Two rational n-knots (Σ, K) and (Σ', K') are called *(rationally) concordant* if there is a rational homology cobordism W between Σ and Σ' and a properly embedded $S^n \times [0,1]$ in W which is bounded by $K \cup -K'$. If (Σ, K) is rationally concordant to the unknot in S^{n+2}, we call it a *(rational) slice knot*.

When K and K' are knots in $S^{n+2} \times 0$ and $S^{n+2} \times 1$ and W is $S^{n+2} \times [0,1]$ in Definition 2.2, it becomes the definition of ordinary concordance. In this case we sometimes say that K and K' are *integrally concordant*, to distinguish it from rational concordance.

From now on, a "knot" means a rational knot unless it is clear from the context that it is an integral knot, and similarly for "concordance".

We denote the set of concordance classes of n-knots by $\mathcal{C}_n$, and denote the set of integral concordance classes of integral n-knots by $\mathcal{C}_n^{\mathbb{Z}}$. $\mathcal{C}_n$ shares some basic properties with $\mathcal{C}_n^{\mathbb{Z}}$; $\mathcal{C}_n$ is an abelian group under the addition and the inversion operations given by connected sum of pairs $(\Sigma, K)\#(\Sigma', K') =$

($\Sigma\#\Sigma', K\#K'$) and orientation reversing $-(\Sigma, K) = (-\Sigma, -K)$, respectively. The identity of $\mathcal{C}_n$ is the concordance class of (rational) slice knots. An n-knot (Σ, K) is in this class if and only if there is a rational $(n+3)$-ball Δ bounded by Σ and a properly embedded $(n+1)$-disk in Δ bounded by K.

DEFINITION 2.3. For a knot K, a codimension one submanifold bounded by K in the ambient space is called a *Seifert surface*.

For an integral knot, there always exists a Seifert surface by a transversality argument. Seifert surfaces play an important role in the study of concordance; for example, Kervaire [26] and Levine [33, 32] computed the structure of $\mathcal{C}_n^{\mathbb{Z}}$ for $n > 1$ using Seifert surfaces.

The first remarkable difference of the rational and integral theories is that a rational knot may not admit any Seifert surface. This leads us to consider a generalized notion of a Seifert surface. Precisely, we adopt the following definition of [7].

DEFINITION 2.4. For an n-knot K in Σ, an $(n+1)$-submanifold F in Σ is called a *generalized Seifert surface* if F is bounded by the union of c disjoint parallel copies of K which are taken along a framing of K (i.e., a trivialization of the normal bundle) for some positive integer c. For $n = 1$, we require that F induces a framing on K which is fiber homotopic to the framing used in taking parallel copies. c is called the *complexity* of F.

We remark that in dimension three, the framing requirement is crucial in later results, and it seems the minimal condition required to extract concordance invariants. For $n > 1$, we do not need the framing condition since any framings of K are fiber homotopic.

EXAMPLE 2.5. Consider the lens space $\Sigma = L(2, 1)$ obtained by $(2/1)$-surgery along an unknotted circle C in S^3. The meridian of C can be viewed as a rational knot K in Σ. In Figure 1, a surface bounded by C and two parallel copies K_1, K_2 of K is illustrated. It can be seen that the surface induces the $(2/1)$-framing on C. Thus, by attaching a disk along a parallel copy of C, we obtain a surface F in Σ which is bounded by $K_1 \cup K_2$. However, according to the definition above, F is *not* a generalized Seifert surface since F gives rise to the $(1/1)$-framing on the K_i while the K_i are taken along the $(0/1)$-framing.

In fact, it turns out that K does not admit any generalized Seifert surface by appealing to Theorem 2.6 below.

In [7], it was shown exactly when a generalized Seifert surface exists. The result in [7] was stated and proved for the more general case of links. For our purpose, it suffices to consider knots only, and in this case, the result can be described in a simpler way as follows.

THEOREM 2.6 ([7]).
(1) *For $n > 1$, any knot admits a generalized Seifert surface.*

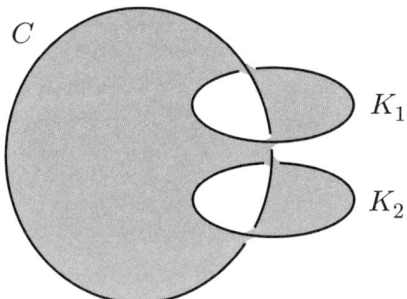

FIGURE 1

(2) *For $n = 1$, a knot admits a generalized Seifert surface if and only if its $\mathbb{Q}/\mathbb{Z}$-valued self-linking in the ambient space is trivial.*

When $n = 1$, it can be easily seen that the $\mathbb{Q}/\mathbb{Z}$-valued self-linking is a concordance invariant of rational knots. From Theorem 2.6 (2), it follows that a knot admits a generalized Seifert surface if and only if so does every knot in the same concordance class. Denoting the subgroup of concordance classes of 1-knots admitting generalized Seifert surfaces by $s\mathcal{C}_1$, we have the following consequence:

COROLLARY 2.7. *$s\mathcal{C}_1$ fits into a short exact sequence*
$$0 \longrightarrow s\mathcal{C}_1 \longrightarrow \mathcal{C}_1 \longrightarrow \mathbb{Q}/\mathbb{Z} \longrightarrow 0.$$

PROOF. By Theorem 2.6 (2), it suffices to show that the self-linking map $\mathcal{C}_1 \to \mathbb{Q}/\mathbb{Z}$ is surjective. For this purpose we generalize the construction of Example 2.5. For an arbitrary positive integer r, consider the lens space Σ obtained by $(r/1)$-surgery along a component of a Hopf link in S^3, and view the other component as a knot K in the lens space. In a similar way as Example 2.5, we can construct a surface F in Σ such that ∂F consists of r parallel copies of K and $F \cap K$ consists of a single point. Thus the self-linking of K is $(1/r) + \mathbb{Z} \in \mathbb{Q}/\mathbb{Z}$. □

Since we need to use the argument of the proof of Theorem 2.6 in the next section as well, we will give a formal proof which is specialized for knots. Another reason to give the proof is that it illustrates the role of the notion of "complexity" of a codimension two pair, which is crucially important in the study of rational concordance. We start with a definition. Suppose M is a codimension two connected submanifold properly embedded in W such that $H_1(W - M; \mathbb{Q})$ is a one-dimensional vector space generated by a meridian of M.

DEFINITION 2.8. The *complexity* of (W, M) (or simply M) is defined to be the absolute value of the element represented by a meridian of M in $H_1(W - M; \mathbb{Z})/\text{torsion} \cong \mathbb{Z}$.

Note that this definition must be distinguished from the complexity of a generalized Seifert surface. The complexity of (W, M) is always a positive integer. As examples to keep in mind, we can think of the complexity when (W, M) is one of the followings:

(1) a knot in a rational sphere,
(2) a slice disk in a rational ball, and
(3) a rational concordance in a rational homology cobordism between rational spheres.

In any case, the Alexander duality shows that (W, M) has the above property.

Assuming that M is framed in W, we identify its regular neighborhood with $M \times D^2$ and denote the exterior $W - \operatorname{int}(M \times D^2)$ by E_M. In particular, $M \times S^1$ is identified with a subspace of ∂E_M. For a space X, we denote by p_X^c the composition
$$p_X^c \colon X \times S^1 \longrightarrow S^1 \longrightarrow S^1$$
of the projection onto S^1 and the map on S^1 given by $z \to z^c$, where S^1 is viewed as the unit circle in the complex plane.

DEFINITION 2.9. $f \colon E_M \to S^1$ is called an S^1-*structure of complexity* c if the composition
$$M \times S^1 \hookrightarrow E_M \xrightarrow{f} S^1$$
is equal to $p_M^c \colon M \times S^1 \to S^1$.

For simplicity we assume that both ∂M and ∂W are connected (or empty). Viewing ∂M as a framed submanifold of ∂W, we assume that $(\partial W, \partial M)$ satisfies our assumption above so that its complexity is defined. Denote its exterior by $E_{\partial M}$.

LEMMA 2.10. *Suppose that an S^1-structure $f \colon E_{\partial M} \to S^1$ of complexity c is given and the homomorphism $H_1(M) \to H_1(E_M)/\text{torsion}$ induced by*
$$M \times \{pt\} \longrightarrow M \times S^1 \longrightarrow E_M$$
is a zero homomorphism. If c is a multiple of the complexity of M, then f extends to an S^1-structure $E_M \to S^1$ of complexity c.

PROOF. Define a map
$$f \cup p_M^c \colon E_{\partial M} \underset{\partial M \times S^1}{\cup} (M \times S^1) \longrightarrow S^1$$
by glueing $f \colon E_{\partial M} \to S^1$ and the map $p_M^c \colon M \times S^1 \to S^1$. We will extend this map to obtain a desired S^1-structure $E_M \to S^1$.

The induced map f_* on H_1 factors through the torsion-free quotient as follows:
$$f_* \colon H_1(E_{\partial M}) \longrightarrow H_1(E_{\partial M})/\text{torsion} = \mathbb{Z} \xrightarrow{\alpha} \mathbb{Z}$$
where α sends a meridian to c since f has complexity c. By the definition of the complexity, there is a unique homomorphism
$$\beta \colon H_1(E_M)/\text{torsion} = \mathbb{Z} \longrightarrow \mathbb{Z}$$

sending a meridian of M to c.

By obstruction theory, it suffices to show that the diagram below commutes:

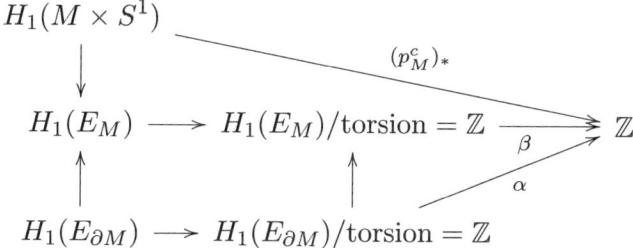

where the vertical homomorphisms are induced by inclusions. The commutativity of the upper triangle can be seen by considering the definition of p_M^c and the following facts: β sends a meridian to c and $H_1(M \times S^1) \to H_1(E_M)/\text{torsion}$ kills cycles from $M \times \{pt\}$. On the other hand, the lower triangle commutes since both α and β send meridians to c and meridians are nonzero in $H_1(E_{\partial M})/\text{torsion}$ and $H_1(E_M)/\text{torsion}$. □

Now we apply Lemma 2.10 to the case of rational knots to show the existence of generalized Seifert surfaces.

PROOF OF THEOREM 2.6. Let K be an n-knot in a rational $(n+2)$-sphere Σ and let E be its exterior.

Suppose $n > 1$ and c is a nonzero multiple of the complexity of K. K always has a trivial normal bundle in Σ and all framings are equivalent so that we can view K as a framed submanifold in a canonical way. Now, since $H_1(K) = 0$, we can appeal to Lemma 2.10 to obtain an S^1-structure $f \colon E \to S^1$ of complexity c. (Note that K and Σ are both closed so that we do not need any maps on the boundary.) By a transversality argument, we can pick a regular value of f whose pre-image is a submanifold F in E. F is a generalized Seifert surface of complexity c.

For $n = 1$ we remark that, although K has a trivial normal bundle, the above argument does not work since framings are not unique and $H_1(K)$ is nontrivial. Indeed this gives rise to our self-linking obstruction. First we show the necessity of the vanishing of the obstruction. If there is a generalized Seifert surface F of complexity c, then the $\mathbb{Q}/\mathbb{Z}$-valued self-linking of K is equal to, modulo $\mathbb{Z}$, $(1/c)(F \cdot K)$ where $\cdot$ denotes the intersection number in Σ. By the framing condition in the definition of a generalized Seifert surface, $F \cdot K = 0$. Thus the linking is trivial.

For the converse, suppose that the $\mathbb{Q}/\mathbb{Z}$-valued self-linking is trivial. Identify ∂E with $K \times S^1$ by choosing a framing on K, and let $\lambda = K \times \{pt\}$ and $\mu = \{pt\} \times S^1$ be the associated longitude and meridian, respectively. Choose a 2-chain u in Σ bounded by $r\lambda$ ($r \neq 0$) such that u meets K transversally. The self-linking of K is, modulo $\mathbb{Z}$, $(1/r)(u \cdot K)$. Since it vanishes, $u \cdot K = kr$ for some integer k. This shows that $r(\lambda + k\mu) = 0$ in $H_1(E)$. Since $\lambda + k\mu$ is the preferred longitude of some framing, we may assume that $r\lambda = 0$ in $H_1(E)$ by changing the framing. Therefore $H_1(K \times \{pt\}) \to H_1(E)/\text{torsion}$ is trivial. Now by using Lemma 2.10 and a transversality argument as above, we can produce a generalized Seifert surface of complexity c for any multiple c of the complexity of K. □

REMARK 2.11. The proof of Theorem 2.6 shows that an n-knot K admits a generalized Seifert surface of complexity c if and only if c is a multiple of the complexity of K (and in addition, for $n = 1$, K has vanishing $\mathbb{Q}/\mathbb{Z}$-valued self-linking).

2.2. Limits of Seifert matrices

Now we focus on odd dimensional knots. For notational convenience, let $n = 2q - 1$ and $\epsilon = (-1)^{q+1}$. We recall some fundamental results about integral concordance: for an integral knot with a Seifert surface F, the Seifert pairing
$$S \colon H_q(F; \mathbb{Z}) \times H_q(F; \mathbb{Z}) \longrightarrow \mathbb{Z}$$
is defined by $S(x, y) = \text{lk}(x^+, y)$ for q-cycles x and y on F where x^+ denotes the cycle obtained by pushing x slightly along the positive normal direction of F. Choosing a basis of $H_q(F; \mathbb{Z})$ modulo the torsion subgroup, A matrix A is associated to S. A is called a *(integral) Seifert matrix*. An integral Seifert matrix A has the property that $A - \epsilon A^T$ is unimodular (i.e., invertible over $\mathbb{Z}$) where A^T designates the transpose of A. In fact this characterizes integral Seifert matrices; a square integral matrix A with this property is a Seifert matrix of an integral knot.

DEFINITION 2.12. A square matrix A is called *metabolic* if it is of even dimension, say $2g$, and congruent to a matrix whose top-upper $g \times g$ submatrix is zero. Two square matrices A and B are called *cobordant* if the block sum $A \oplus (-B)$ is a metabolic matrix.

In [**33**], Levine proved that cobordism is an equivalence relation of integral Seifert matrices, and their equivalence classes form an abelian group under block sum, which is called the *algebraic concordance group*. We denote it by $G_n^{\mathbb{Z}}$. He also proved that Seifert matrices of integrally concordant knots are cobordant. This establishes a group homomorphism $\mathcal{C}_n^{\mathbb{Z}} \to G_n^{\mathbb{Z}}$. The following result of Levine transforms the geometric problem of (integral) knot concordance in higher dimensions into an equivalent algebraic problem:

THEOREM 2.13 (Levine [**33**]). $\mathcal{C}_n^{\mathbb{Z}} \to G_n^{\mathbb{Z}}$ is an isomorphism for odd $n > 3$, an injection onto an index two subgroup of $G_n^{\mathbb{Z}}$ for $n = 3$, and a surjection for $n = 1$.

Furthermore Levine computed $G_n^{\mathbb{Z}}$ using associated isometric structures.

THEOREM 2.14 (Levine [**32**]). $G_n^{\mathbb{Z}} \cong \mathbb{Z}^\infty \oplus (\mathbb{Z}/2)^\infty \oplus (\mathbb{Z}/4)^\infty$.

It can be seen that the image of $\mathcal{C}_3^{\mathbb{Z}} \to G_3^{\mathbb{Z}}$ is abstractly isomorphic to the same group too. This gives us a full calculation of the integral knot concordance group for $n > 1$.

Returning to the discussion of rational knots, a Seifert matrix of a rational knot can be defined in a similar way, using generalized Seifert surfaces. (However, as we will see later, a rational analogue of the algebraic concordance group $G_n^{\mathbb{Z}}$ is constructed in a more sophisticated way.) For this purpose we need the rational-valued linking number, which is a straightforward generalization of the integral linking number in S^{n+2}. For concreteness, we give a definition below.

DEFINITION 2.15. Let x and y be disjoint p-cycle and q-cycle in a rational $(p+q+1)$-sphere Σ, respectively. Then the *linking number* of x and y in Σ is defined to be $\mathrm{lk}_\Sigma(x,y) = (1/b)(x \cdot v)$, where v is a $(q+1)$-chain bounded by by for some $b \neq 0$.

It is straightforward to check that the linking number is well-defined.

REMARK 2.16.
(1) The rational-valued linking number is defined for disjoint cycles only, and not well-defined on the homology classes; its modulo $\mathbb{Z}$ reduction is the usual $\mathbb{Q}/\mathbb{Z}$-valued linking which is well-defined for homology classes.
(2) If x is a connected submanifold of dimension p which is embedded in Σ, then $H_q(\Sigma - x; \mathbb{Q})$ can be identified with $\mathbb{Q}$ in such a way that for any q-cycle y in $\Sigma - x$, the linking number of x and y is the element $[y]$ in $H_q(\Sigma - x; \mathbb{Q}) = \mathbb{Q}$. In particular, a meridian of x corresponds to $1 \in \mathbb{Q}$.
(3) For a computation formula of the linking number from a surgery description of Σ, see [7].

Suppose K is an n-knot admitting a generalized Seifert surface F. Now a bilinear pairing over $\mathbb{Q}$
$$S \colon H_q(F; \mathbb{Q}) \times H_q(F; \mathbb{Q}) \longrightarrow \mathbb{Q}$$
can be defined by the same formula as the integral Seifert pairing, using the rational-valued linking number.

For $n = 1$, it turns out that homology classes from boundary components of F have no interesting information.

LEMMA 2.17. *For a generalized Seifert surface F of a 1-knot, $S(x,y)$ vanishes if either x or y is a cycle from ∂F.*

PROOF. Suppose y is a component of ∂F. Then F can be viewed as a 2-chain bounded by cy, where $c \neq 0$ is the complexity of F. Since F is disjoint from x^+ for any 1-cycle x on F,
$$S(x,y) = (1/c)(x^+ \cdot F) = 0.$$
The same argument also works when x is from ∂F. □

From this it follows that S gives rise to a well-defined pairing on the cokernel of $H_1(\partial F; \mathbb{Q}) \to H_1(F; \mathbb{Q})$.

DEFINITION 2.18. S is called the *(rational-valued) Seifert pairing* of F. For $n > 1$, a matrix associated to S by choosing a basis of $H_q(F; \mathbb{Q})$ is called a *(rational) Seifert matrix of complexity c*, where c is the complexity of F. For $n = 1$, a matrix associated to the induced pairing on the cokernel of $H_1(\partial F; \mathbb{Q}) \to H_1(F; \mathbb{Q})$ is called a *(rational) Seifert matrix of complexity c*.

A rational Seifert matrix has the following property which is analogous to the characterization property of integral Seifert matrices:

LEMMA 2.19. *If A is a Seifert matrix, then for some rational square matrix P, $P(A - \epsilon A^T)P^T$ is integral and even unimodular over $\mathbb{Z}$.*

Here "even" means that all diagonal entries are even. Lemma 2.19 is an immediate consequence of the fact that $A - \epsilon A^T$ represents the rational intersection form on $H_q(F; \mathbb{Q})$, which is obtained from the integral intersection form on $H_q(F; \mathbb{Z})$ by tensoring $\mathbb{Q}$.

REMARK 2.20. In Section 4.1, we will show that the property described in Lemma 2.19 is a characterization of a (rational) Seifert matrix. In fact, in Theorem 4.1, we give a constructive realization: for a matrix A having the property in Lemma 2.19 and an arbitrary positive integer c, there is a knot which has A as a Seifert matrix of complexity c.

Now we use rational Seifert matrices to construct an abelian group $\mathcal{G}_n$ which can be viewed as a "rationalization" of $G_n^{\mathbb{Z}}$. We will also construct group homomorphisms $\Phi_n \colon \mathcal{C}_n \to \mathcal{G}_n$ for odd $n > 1$ and $\Phi_1 \colon s\mathcal{C}_1 \to \mathcal{G}_1$ for $n = 1$, which are analogous to the Levine homomorphism $\mathcal{C}_n^{\mathbb{Z}} \to G_n^{\mathbb{Z}}$. Indeed $\mathcal{G}_n$ is a limit of Levine's ordinary algebraic cobordism groups of matrices. As before, we continue to use the convention $n = 2q - 1$ and $\epsilon = (-1)^{q+1}$. Consider the set of all rational square matrices A having the property in Lemma 2.19. As in the case of integral Seifert matrices, an argument in [32] shows that matrix cobordism is an equivalence relation on this set, and the set of cobordism classes of matrices with the property in Lemma 2.19 becomes an abelian group under the block sum. We denote this group by $\mathcal{G}_n$. The cobordism class of A will be denoted by $[A]$.

Sometimes we call a matrix A with the property in Lemma 2.19 a *(rational) Seifert matrix*, as an abuse of terminology at this time. As mentioned in Remark 2.20, it will be justified later.

REMARK 2.21. In [7], rational Seifert matrices were viewed as representatives of elements of another group $G_\epsilon^{\mathbb{Q}}$; they considered all square matrices A such that $A - \epsilon A^T$ is nonsingular, instead of the property in Lemma 2.19, and formed the group $G_\epsilon^{\mathbb{Q}}$ of cobordism classes of such matrices. For odd q (i.e., $\epsilon = +1$), it turns out that $G_\epsilon^{\mathbb{Q}}$ coincides with $\mathcal{G}_n$. For, if $A - A^T$ is nonsingular, then it is congruent to a block sum of 2 by 2 matrices $\begin{bmatrix} 0 & 1 \\ -1 & 0 \end{bmatrix}$ since it is skew-symmetric. For even q, however, $\mathcal{G}_n$ is a proper subgroup of $G_\epsilon^{\mathbb{Q}}$. A way to see this fact is to observe the following property: if $[A] \in \mathcal{G}_n$, then $\det(A + A^T)$ is a square in $\mathbb{Q}$. This condition is not necessarily satisfied by elements in $G_\epsilon^\mathbb{Q}$. For example, for $A = \begin{bmatrix} 1 & 3 \\ 0 & 1 \end{bmatrix}$, $[A]$ is an element of $G_\epsilon^\mathbb{Q}$ which is not in G_n.

For a square matrix A, we denote by $i_r A$ the matrix
$$\begin{bmatrix} A & A & A & \cdots & A \\ \epsilon A^T & A & A & \cdots & A \\ \epsilon A^T & \epsilon A^T & A & \cdots & A \\ \vdots & \vdots & \vdots & \ddots & \vdots \\ \epsilon A^T & \epsilon A^T & \epsilon A^T & \cdots & A \end{bmatrix}$$
consisting of $r \times r$ blocks (submatrices below the diagonal blocks are ϵA^T, and all the other submatrices are A). Then $A \to i_r A$ gives rise to an endomorphism on G_n, which we also denote by i_r. Note that if A is a Seifert matrix of a generalized Seifert surface F, $i_r A$ is a Seifert matrix of the union of r parallel copies of F.

Let $G_{n,c} = G_n$ for each positive integer c, and let $\phi_{c,d} \colon G_{n,c} \to G_{n,d}$ be $i_{d/c}$ for every pair (c,d) of positive integers such that $c \mid d$. Then $(\{G_{n,c}\}, \{\phi_{c,d}\})$ becomes a direct system.

DEFINITION 2.22. The limit $\mathcal{G}_n = \varinjlim_c G_{n,c}$ is called the *algebraic rational concordance group*.

We denote the natural homomorphism $G_{n,c} \to \mathcal{G}_n$ by ϕ_c.

For $n > 1$, we define $\Phi_n \colon \mathcal{C}_n \to \mathcal{G}_n$ as follows. For any n-knot K, there is a Seifert matrix A of complexity c for some $c > 0$. The image of the concordant class of K under Φ_n is defined to be $\phi_c[A]$, i.e., the image of $[A] \in G_n = G_{n,c}$ under $\phi_c \colon G_{n,c} \to \mathcal{G}_n$. For $n = 1$, we define Φ_1 to be a homomorphism of $s\mathcal{C}_1$; since a 1-knot representing an element in $s\mathcal{C}_1$ has a generalized Seifert surface, we can associate an element of $\mathcal{G}_n$ in the same way.

THEOREM 2.23. Φ_n *is a well-defined group homomorphism.*

PROOF. First we prove the additivity. Suppose K_1 and K_2 are knots with Seifert matrices A_1 and A_2 of complexity c_1 and c_2, respectively. Then $i_{c_2} A_1$ and $i_{c_1} A_2$ have the same complexity $c_1 c_2$. For $n > 1$, by a Mayer–Vietoris argument it is easily seen that $A = i_{c_2} A_1 \oplus i_{c_1} A_2$ is a Seifert matrix of complexity $c_1 c_2$ for $K_1 \# K_2$. For $n = 1$, although it might not be true, arguments in the proof of [**7**, Theorem 1.2 (4)] show that $i_{c_2} A_1 \oplus i_{c_1} A_2$ is *cobordant* to a Seifert matrix A of complexity $c_1 c_2$ for $K_1 \# K_2$. It implies the desired additivity: $\phi_{c_1}[A_1] + \phi_{c_2}[A_2] = \phi_{c_1 c_2}[A]$ in $\mathcal{G}_n$.

Now it suffices to show that Φ_n is well-defined for rational slice knots. Once it is proved, general well-definedness follows from the additivity. We give a unified proof for any odd n. Suppose that Δ is a rational $(n+3)$-ball with boundary Σ, and D is a properly embedded $(n+1)$-disk in Δ whose boundary is a knot K in Σ. Suppose F is a generalized Seifert surface of complexity c for K and A is a Seifert matrix defined on F. Note that D

has a unique framing in Δ and its restriction on K agrees with the framing induced by F. Choose a common multiple r of c and the complexity of (Δ, D). Taking r/c parallel copies of F and applying a Thom–Pontryagin construction, we obtain an S^1-structure $f\colon E_K \to S^1$ of complexity r such that $f^{-1}(pt)$ is a generalized Seifert surface with Seifert matrix $i_{r/c}A$. Since $H_1(D) = 0$, we can apply Lemma 2.10 to obtain an S^1-structure $g\colon E_D \to S^1$ which extends f. By a transversality argument for g, we construct a $(n+2)$-submanifold R in Δ such that

$$\partial R = (r \text{ parallel copies of } D) \underset{\partial}{\cup} (r/c \text{ parallel copies of } F).$$

$H_q(\partial R; \mathbb{Q})$ is isomorphic to the direct sum of r/c copies of

$$\operatorname{Coker}\{H_q(\partial F; \mathbb{Q}) \longrightarrow H_q(F; \mathbb{Q})\}$$

which is equal to $H_q(F; \mathbb{Q})$ for $n > 1$, and $i_{r/c}A$ represents a bilinear pairing on $H_q(\partial R; \mathbb{Q})$, for any odd n including $n = 1$. As in [**33**],

$$\operatorname{Ker}\{H_q(\partial R; \mathbb{Q}) \longrightarrow H_q(R; \mathbb{Q})\}$$

is a half-dimensional subspace on which this pairing vanishes. Thus $i_{r/c}A$ is metabolic. It completes the proof. $\square$

The above proof shows that, if A is a Seifert matrix of complexity c for a knot K which admits a slice disk of complexity c' in a rational ball, then for any common multiple r of c and c', $i_{r/c}A$ is metabolic. For later use, we state an analogue for concordant knots:

COROLLARY 2.24 (Corollary to the proof of Theorem 2.23). *Suppose that K_1 and K_2 are concordant via a concordance of complexity c'. If A_1 and A_2 are Seifert matrices of complexity c_1 and c_2 for K_1 and K_2, respectively, then $i_{r/c_1}A_1$ and $i_{r/c_2}A_2$ are cobordant for any common multiple r of c_1, c_2, and c'.*

PROOF. Let C be a concordance between K_1 and K_2 in a rational homology cobordism W between their ambient spaces. Choosing an arc γ on C joining K_1 and K_2 and removing a regular neighborhood of γ from (W, C), we obtain a pair (Δ, D) of a rational ball Δ and a slice disk D of $K_1 \# (-K_2)$. Since $E_C \cong E_D$, the complexity of D is c'. Now, $i_{r/c_1}A_1 \oplus i_{r/c_2}A_2$ is (cobordant to) a Seifert matrix of complexity r for $K_1 \# (-K_2)$, and from the proof of Theorem 2.23, it follows that $i_{r/c_1}A_1 \oplus (-i_{r/c_2}A_2)$ is metabolic. $\square$

CHAPTER 3

Algebraic structure of $\mathcal{G}_n$

3.1. Invariants of Seifert matrices

In this section we discuss some known invariants of G_n. Levine [**32**] first revealed the structure of G_n using invariants of isometric structures associated to Seifert matrices. We will follow another approach using relative Witt groups of linking forms.

We begin by recalling the definition of the relative Witt group $W_\epsilon(R,S)$ (as a general reference, refer to Ranicki's book [**40**]; see also Hillman's book [**20**]). Let R be a commutative ring with an involution $r \to \bar{r}$, S be a multiplicative subset in R, and $\epsilon = \pm 1$. $S^{-1}R/R$ has an induced involution which will also be denoted by the same notation. If a sesquilinear map $B\colon V\times V \to S^{-1}R/R$ on a finitely generated S-torsion R-module V is nonsingular and ϵ-hermitian, then we call it an *ϵ-linking pairing over (R,S)*. Here ϵ-hermitian means that $B(x,y) = \epsilon \overline{B(y,x)}$, and nonsingular means that the adjoint map $V \to \operatorname{Hom}(V, S^{-1}R/R)$ is bijective. (For our purpose, the usual homological dimension condition is not needed.) V together with B is called an *ϵ-linking form* over (R,S). As an abuse of notation, we sometimes denote it simply by B. The direct sum of two ϵ-linking forms are defined in an obvious way.

An ϵ-linking form $B\colon V\times V \to S^{-1}R$ is said to be *hyperbolic* if there is a submodule $P \subset V$ such that
$$P^\perp = \{x \in V \mid B(x,y) = 0 \text{ for all } y \in P\}$$
is equal to P. Two ϵ-linking forms B_1 and B_2 are said to be *Witt equivalent* if $B_1 \oplus B' \cong B_2 \oplus B''$ for some hyperbolic ϵ-linking forms B' and B''. It is an equivalence relation, and the equivalence classes form an abelian group under the direct sum operation. It is called the *relative Witt group* $W_\epsilon(R,S)$.

For the study of the algebraic concordance group G_n, first we relate G_n with a particular relative Witt group $W_\epsilon(\mathbb{Q}[t^{\pm 1}], S)$, where $\mathbb{Q}[t^{\pm 1}]$ is the Laurent polynomial ring equipped with the involution $\bar{t} = t^{-1}$ and S is the multiplicative subset of all nonzero elements in $\mathbb{Q}[t^{\pm 1}]$. Then, the structure of $W_\epsilon(\mathbb{Q}[t^{\pm 1}], S)$ is well-understood via "devissage", using the key advantage of this case that $\mathbb{Q}[t^{\pm 1}]$ is a PID.

DEFINITION 3.1. A polynomial $\lambda(t)$ in $\mathbb{Q}[t^{\pm 1}]$ is called *reciprocal* if $\lambda(t) = u\lambda(t^{-1})$ for some unit u in $\mathbb{Q}[t^{\pm 1}]$. An algebraic number z is called *reciprocal* if its irreducible polynomial is reciprocal.

A prime ideal in $\mathbb{Q}[t^{\pm 1}]$ is preserved by the involution if and only if it is generated by a reciprocal irreducible polynomial $\lambda(t)$. In this case, $\mathbb{Q}[t^{\pm 1}]/\langle\lambda(t)\rangle$ becomes a field with an induced involution. Then we can think of the ordinary Witt group $W_\epsilon(\mathbb{Q}[t^{\pm 1}]/\langle\lambda(t)\rangle)$ of nonsingular ϵ-hermitian forms $b\colon V \times V \to \mathbb{Q}[t^{\pm 1}]$ on finite dimensional vector spaces V over the field $\mathbb{Q}[t^{\pm 1}]/\langle\lambda(t)\rangle$.

PROPOSITION 3.2. *There are injective homomorphisms*
$$\mathcal{G}_n \longrightarrow W_\epsilon(\mathbb{Q}[t^{\pm 1}], S) \longrightarrow \bigoplus W_\epsilon(\mathbb{Q}[t^{\pm 1}]/\langle\lambda(t)\rangle)$$
where the sum is taken over all reciprocal irreducible polynomials $\lambda(t)$.

For later use, we briefly describe the homomorphisms. For a representative A of an element of $\mathcal{G}_n$, the $\mathbb{Q}[t^{\pm 1}]$-module V presented by the matrix $tA - \epsilon A^T$ is a torsion module since its determinant is a polynomial whose evaluation at $t = 1$ is $\det(A - \epsilon A^T) \ne 0$. The matrix $(1-t)(tA - \epsilon A^T)^{-1}$ gives rise to an ϵ-linking pairing B on V. (V, B) is called the *Blanchfield form associated to A*. It is straightforward to verify that this gives rise to a well-defined group homomorphism $\mathcal{G}_n \to W_\epsilon(\mathbb{Q}[t^{\pm 1}], S)$. This is the first homomorphism in Proposition 3.2. Kearton showed an analogous homomorphism of $\mathcal{G}_n^\mathbb{Z}$ is injective [24]. Although he considered integral Seifert matrices only, exactly the same argument works for rational Seifert matrices as well. We do not repeat the details.

The second homomorphism in Proposition 3.2 is described as follows. Let B be an ϵ-linking form on a torsion $\mathbb{Q}[t^{\pm 1}]$-module V. Since $\mathbb{Q}[t^{\pm 1}]$ is a PID, V is canonically decomposed into the direct sum of primary subspaces: $V = \bigoplus V_{\lambda(t)}$ where
$$V_{\lambda(t)} = \{v \in V \mid \lambda(t)^N v = 0 \text{ for some } N\}$$
and $\lambda(t)$ runs over reciprocal irreducible polynomials. A standard argument shows that the restriction $B|_{V_{\lambda(t)}}$ is Witt equivalent to a linking form $B_{\lambda(t)}$ on a $\mathbb{Q}[t^{\pm 1}]$-module annihilated by $\lambda(t)$ (e.g., see [32] or [20, page 131]). Then $B_{\lambda(t)}$ can be viewed as an ϵ-hermitian pairing over $\mathbb{Q}[t^{\pm 1}]/\langle\lambda(t)\rangle$. This gives rise to the desired homomorphism. A devissage argument shows that it is injective. For a detailed proof, see [40] or [20].

REMARK 3.3. It is well known that the Blanchfield form can be described geometrically; For an n-knot with $n = 2q - 1$, a generalized Seifert surface induces an S^1-structure via the Thom–Pontryagin construction, which gives us a $\mathbb{Q}[t^{\pm 1}]$ local coefficient system on the knot exterior E_K. The homology module $H_q(E_K; \mathbb{Q}[t^{\pm 1}])$ is presented by $tA - \epsilon A^T$, and $(t-1)(tA - \epsilon A^T)^{-1}$ is known to be the torsion linking pairing on $H_q(E_K; \mathbb{Q}[t^{\pm 1}])$ due to Blanchfield [1]. In [12], the algebraic rational concordance group was defined in terms of this geometric linking form.

Sometimes it is convenient to parametrize the above number fields using reciprocal numbers instead of polynomials. For a zero z of $\lambda(t)$, we identify

$\mathbb{Q}[t^{\pm 1}]/\langle\lambda(t)\rangle$ with $\mathbb{Q}(z)$ via $t \to z$. Given a Seifert matrix, we obtain an associated Witt class of a hermitian form over $\mathbb{Q}[t^{\pm 1}]/\langle\lambda(t)\rangle = \mathbb{Q}(z)$ via the composite map in Proposition 3.2. We call it the $\lambda(t)$-*primary part* or z-*primary part*.

Now we define invariants of G_n via $W_\epsilon(\mathbb{Q}(z))$. The scalar multiplication by $(z - z^{-1})$ induces an injection

$$W_-(\mathbb{Q}(z)) \longrightarrow W_+(\mathbb{Q}(z))$$

for $z \neq \pm 1$, and $W_-(\mathbb{Q}(\pm 1)) = W_-(\mathbb{Q})$ is trivial since any skew-hermitian form over $\mathbb{Q}$ is hyperbolic and so Witt trivial. So we focus on $W_+(\mathbb{Q}(z))$. For a Witt class $[b]$ in $W_+(\mathbb{Q}(z))$ which is represented by a hermitian form b, let r be the $\mathbb{Q}(z)$-dimension of the underlying vector space of b. Then we define rank$[b] \in \mathbb{Z}/2$ to be the residue class of r modulo 2 and dis$[b]$ to be the discriminant

$$\mathrm{dis}[b] = (-1)^{r(r+1)/2} \det b \in \frac{\mathbb{Q}(z + z^{-1})^\times}{N_z^\times}$$

where

$$N_z^\times = \{u\bar{u} \mid u \in \mathbb{Q}(z)^\times\}.$$

For $|z| = 1$, we can also define the signature sign$[b]$ by viewing b as a complex hermitian form via the embedding $\mathbb{Q}(z) \subset \mathbb{C}$. Then it is known that they are well-defined, and furthermore,

PROPOSITION 3.4 ([36]). *If $z \neq \pm 1$ is reciprocal, then* sign, rank, *and* dis *form a complete set of invariants of* $W_+(\mathbb{Q}(z))$. *In other words,* $[b] \in W_+(\mathbb{Q}(z))$ *is trivial if and only if* sign$[b]$, rank$[b]$, *and* dis$[b]$ *are trivial.*

For $z \neq \pm 1$, we can define analogous invariants sign, rank, and dis of elements in $W_-(\mathbb{Q}(z))$ by composing the above invariants with the injection into $W_+(\mathbb{Q}(z))$. Then these invariants for $W_-(\mathbb{Q}(z))$ are also complete.

Consider $[A] \in G_n$ and a reciprocal number z. In addition if $\epsilon = -1$, suppose that $z \neq \pm 1$. We denote sign A_z, rank A_z, and dis A_z of the z-primary part $A_z \in W_\epsilon(\mathbb{Q}(z))$ by $s_z[A]$, $e_z[A]$, and $d_z[A]$, respectively. For notational convenience, we define $s_z[A]$, $e_z[A]$, and $d_z[A]$ to be trivial when $\epsilon = -1$ and $z = \pm 1$. Recall that A_z is always Witt trivial in this case.

REMARK 3.5. For $z = \pm 1$ and $\epsilon = +1$ (i.e., q is even), the above invariants of $W_+(\mathbb{Q}(z)) = W_+(\mathbb{Q})$ are not complete. Indeed the structure of $W_+(\mathbb{Q})$ is quite different since the involution is trivial. For integral knots, it is known that the (± 1)-primary part of an integral Seifert matrix is always trivial for any q and so we do not need to consider this in investigating the structure of $G_n^\mathbb{Z}$. While elements of our G_n have trivial $(+1)$-primitive part, the (-1)-primary part is not necessarily trivial. In fact, such an example can be produced by using our realization theorem that will be proved later (Theorem 3.22). This shows that our invariants of G_n are not complete. However it will turn out that a complete set of invariants of $\mathcal{G}_n$ can be extracted from these invariants. Hence we do not discuss the structure of $W_+(\mathbb{Q})$ in this paper. Interested readers may refer to Milnor–Husemoller [**36**].

It is known that we can sometimes evaluate the invariants directly from A, without computing the z-primary part explicitly.

PROPOSITION 3.6.
(1) If the irreducible polynomial of z does not appear (i.e., has exponent zero) in the factorization of the Alexander polynomial
$$\Delta_A(t) = \det(tA - \epsilon A^T),$$
then the z-primary part of A is trivial. In particular, $s_z(A)$, $e_z(A)$, and $d_z(A)$ are trivial.
(2) For $|z| = 1$, $s_z[A]$ is the jump of the signature function $S^1 \to \mathbb{Z}$ given by
$$w \longrightarrow \begin{cases} \operatorname{sign} \dfrac{wA - \epsilon A}{w-1} & \text{for } \epsilon = 1 \\ \operatorname{sign}(w - \bar w)\dfrac{wA - \epsilon A}{w-1} & \text{for } \epsilon = -1 \end{cases}$$
at $w = z$.
(3) $e_z[A]$ is congruent, modulo 2, to the exponent of the irreducible polynomial of z in the factorization of the Alexander polynomial $\Delta_A(t)$.
(4) s_z and e_z are additive, i.e.,
$$s_z([A] + [B]) = s_z[A] + s_z[B],$$
$$e_z([A] + [B]) = e_z[A] + e_z[B].$$
For d_z, we have
$$d_z([A] + [B]) \equiv (-1)^{e_z[A]e_z[B]} d_z[A]d_z[B] \mod N_z^\times.$$
In particular, $d_z(2[A]) \equiv (-1)^{e_z[A]} \mod N_z^\times$.

PROOF. Let B be the Blanchfield form of A. Recall that the order of a $\mathbb{Q}[t^{\pm 1}]$-module is defined to be the determinant of a square presentation matrix. The order of the underlying module V of B is equal to $\Delta(t)$ since V is presented by $tA - A^T$. On the other hand, writing the underlying module V of B as a direct sum of cyclic modules $\mathbb{Q}[t^{\pm 1}]/\langle p_i(t)^{n_i}\rangle$, where $p_i(t)$ is irreducible, the order of V is $\prod p_i(t)^{n_i}$. Thus $\Delta_A(t) = \prod p_i(t)^{n_i}$. Furthermore, from the above decomposition of V, we can observe that the primary subspace $V_{\lambda(t)}$ is trivial if $\lambda(t) \neq p_i(t)$ (up to units) for all i. From this (1) follows. (2) was proved in [**35**]. For (3), let $\lambda(t)$ be the irreducible polynomial of z and e be the exponent of $\lambda(t)$ in the factorization of $\Delta_A(t)$. Then $\lambda(t)^e$ is the order of the subspace $V_{\lambda(t)}$ of V. From the fact that the order of the underlying module of a hyperbolic linking pairing is of the form $f(t)f(t^{-1})$, it follows that the modulo 2 residue class of the exponent e is a Witt invariant of B. Thus we may assume that the $V_{\lambda(t)}$ is annihilated by $\lambda(t)$, as in

the discussion below Proposition 3.2. Writing $V_{\lambda(t)} = (\mathbb{Q}[t^{\pm 1}]/\langle\lambda(t)\rangle)^r$, we have
$$\lambda(t)^r = (\text{order of } V_{\lambda(t)}) = \lambda(t)^e.$$
(4) is proved by a straightforward computation based on our definitions. (Note that the value of d_z lives in a multiplicative group.) □

3.2. Invariants of limits of Seifert matrices

This section is devoted to an algebraic study of invariants of $\mathcal{G}_n$. As before, $n = 2q - 1$ and $\epsilon = (-1)^{q+1}$ throughout this section.

Recall that $\mathcal{G}_n$ is the limit of the direct system consisting of $G_{n,c} = G_n$ and the homomorphisms $\phi_{c,rc} = i_r$. The following result, which is called the reparametrization formula, is crucial in understanding the relationship between this direct system and the invariants of G_n which were discussed in the previous section.

LEMMA 3.7. *If z is reciprocal, then z^r is reciprocal for any positive integer r.*

PROOF. First observe that an irreducible polynomial $p(t)$ is reciprocal if and only if $p(w) = 0 = p(w^{-1})$ for some w. Let $\lambda(t)$ and $\mu(t)$ be the irreducible polynomials of the given z and z^r, respectively. Since z is a zero of $\mu(t^r)$, $\mu(t^r)$ is a multiple of $\lambda(t)$. Since $\lambda(t)$ is reciprocal, $\lambda(z^{-1}) = 0$ and so $\mu(z^{-r}) = 0$. It follows that $\mu(t)$ is reciprocal. □

LEMMA 3.8 (Reparametrization formula). *Suppose $[A] \in \mathcal{G}_n$, z is a reciprocal number, and r is a positive integer. Then*

(1) $s_z(i_r[A]) = s_{z^r}[A]$ *for* $|z| = 1$.
(2) $e_z(i_r[A]) = e_{z^r}[A]$.
(3) $d_z(i_r[A]) \equiv d_{z^r}[A] \mod N_z^\times$.

The conclusion for the signature is already known; Cochran and Orr [12] proved the signature formula using the Blanchfield form. Ko and the author gave an algebraic proof of the signature formula using Seifert matrices [7]. As related work in dimension three, see also Kearton [25] and Litherland [34]. Here we give a single unified proof of all the conclusions above including the new parts (2) and (3).

PROOF. First we claim that

$$\begin{array}{ccc} G_n & \longrightarrow & W_\epsilon(\mathbb{Q}[t^{\pm 1}], S) \\ i_r \downarrow & & \downarrow t \to t^r \\ G_n & \longrightarrow & W_\epsilon(\mathbb{Q}[t^{\pm 1}], S) \end{array}$$

is commutative, where the horizontal homomorphisms are the inclusions discussed in the previous section and the left vertical homomorphism is induced by $t \to t^r$. One way to see the commutativity is to appeal to the geometric interpretation of the horizontal homomorphism (Remark 3.3), as follows: $A \to i_r A$ is induced by taking r parallel copies of a generalized Seifert surface, and hence its effect on the $\mathbb{Q}[t^{\pm 1}]$-coefficient system on the knot exterior discussed in Remark 3.3 is exactly $t \to t^r$.

For concreteness, we outline a purely algebraic argument based on our definitions. Suppose A is a $d \times d$ Seifert matrix. First we show that the presentation $t i_r A - \epsilon i_r A^T$ on rd generators is reduced into a new presentation $t^r A - \epsilon A^T$ on d generators by suitable row and column operations. Indeed, let

$$P = \prod_{i=1}^{r-1} \begin{bmatrix} I_{i-1} & & & \\ & \epsilon A^T & I & \\ & A & I & \\ & & & I_{r-i-1} \end{bmatrix},$$

$$Q = \prod_{i=r-1}^{1} \begin{bmatrix} I_{i-1} & & & \\ & -t^i & I & \\ & I & 0 & \\ & & & I_{r-i-1} \end{bmatrix}$$

where I_k is the $kd \times kd$ identity matrix and $I = I_1$ (the products are expanded from left to right). Then we can check that both P and Q are unimodular over $\mathbb{Q}[t^{\pm 1}]$ (here we need that $A - \epsilon A^T$ is nonsingular) and that $P^{-1}(t i_r A - \epsilon i_r A^T) Q^{-1}$ is of the form

$$R = \begin{bmatrix} I & & & \\ * & \ddots & & \\ \vdots & \ddots & I & \\ * & \cdots & * & t^r A - \epsilon A^T \end{bmatrix}.$$

Under the convention that columns of a presentation matrix represent relations, P is our generator changing matrix. Since its bottom-right $d \times d$ block is I, the generators of the new presentation are exactly the last d generators of the old presentation. Thus the linking pairing for $i_r A$ is given by the bottom-right $d \times d$ block of

$$P^T \cdot (t-1)(t i_r A - \epsilon i_r A^T)^{-1} \cdot P = (t-1) P^T Q^{-1} R.$$

By a straightforward calculation we can check that it is equal to

$$(t^r - 1)(t^r A - A^T)^{-1}.$$

This proves the claim.

Now suppose $[b] \in W_\epsilon(\mathbb{Q}[t^{\pm 1}], S)$. Denote its image under $t \to t^r$ by $[i_r b]$. We will compute the z-primary part of $i_r b$. Consider a special case that b itself is a single primary part. Then we can assume that b is defined on a module V annihilated by a reciprocal irreducible polynomial $\lambda(t)$, i.e., $V = (\mathbb{Q}[t^{\pm 1}]/\langle \lambda(t) \rangle)^d$. Let

$$\lambda(t^r) = \mu_1(t) \cdots \mu_m(t)$$

be the factorization of $\lambda(t^r)$ into distinct irreducible factors. Note that there is no repeated factor since $\lambda(t)$ has no multiple root. Then the underlying module of $i_r b$ is

$$(\mathbb{Q}[t^{\pm 1}]/\langle \lambda(t^r) \rangle)^d \cong \bigoplus_{i=1}^{m} (\mathbb{Q}[t^{\pm 1}]/\langle \mu_i(t) \rangle)^d,$$

where its $\mu_i(t)$-primary part is the $(\mathbb{Q}[t^{\pm 1}]/\langle \mu_i(t) \rangle)^d$-summand. Since

$$(\mathbb{Q}[t^{\pm 1}]/\langle \mu_i(t) \rangle)^d \longrightarrow (\mathbb{Q}[t^{\pm 1}]/\langle \lambda(t^r) \rangle)^d$$

is the multiplication by $\prod_{j \neq i} \mu_j(t)$, the $\mu_i(t)$-primary part of $i_r b$ is given by

$$(e_k^i, e_\ell^i) \longrightarrow \left(\prod_{j \neq i} \mu_j(t) \mu_j(t^{-1}) \right) \cdot \left(b(e_k, e_\ell)|_{t \to t^k} \right)$$

where $\{e_k^i\}$ and $\{e_k\}$ are the standard bases of $(\mathbb{Q}(t)/\langle \mu_i(t) \rangle)^d$ and V, respectively.

This shows that the z-primary part of $i_r b$ is nontrivial if and only if z is a zero of $\mu_i(t)$ for some i, or equivalently z^r is a zero of $\lambda(t)$. In this case the above computation shows $e_z(i_r A) = d = e_{z^r}(A)$. Moreover, if B is a matrix over $\mathbb{Q}(z^r)$ representing b (which is the z^r-primary part of itself), $w\bar{w}B$ represents the z-primary part of $i_r b$ where $w = \prod_{j \neq i} \mu_j(z)$. It follows that $s_z(i_r A) = s_{z^r}(A)$ for $|z| = 1$ and $d_z(i_r A) \equiv d_{z^r}(A)$ modulo $N_z^\times$. Note that when $\epsilon = -1$ and $z^r = \pm 1$, b is automatically Witt trivial and hence the desired conclusions are immediate consequences of our convention.

To reduce the general case to the above special case, we can think of each primary part of b instead of b. The only remaining thing we have to check is that two distinct primary parts of b never give rise to the same primary part of $i_r b$. Indeed, if $\lambda_1(t)$ and $\lambda_2(t)$ are irreducible polynomials such that $\lambda_1(t^r)$ and $\lambda_2(t^r)$ have a common factor $\mu(t)$, then $\lambda_1(t)$ and $\lambda_2(t)$ have a common root and hence $\lambda_1(t) = \lambda_2(t)$ up to units. It completes the proof. □

REMARK 3.9. The argument of the proof of Lemma 3.8 also proves another version the reparametrization formula for the Alexander polynomial: $\Delta_{i_r A}(t) = \Delta_A(t^r)$ up to multiplication by units in $\mathbb{Q}[t^{\pm 1}]$ [**7**].

REMARK 3.10. In [**12**, page 532], a weaker conclusion on signatures was stated and shown: $s_z(i_r A) = \pm s_{z^r}(A)$. The sign ambiguity was introduced by a typo in their proof; the $\bar{w}$ factor in the matrix representation $w\bar{w}B$ of the z-primary part of $i_r b$ was missing.

Now we define our invariants of $\mathcal{G}_n$. Roughly speaking, one can view the reparametrization formula as a contravariant naturality of the invariants of $G_{n,c}$: the direct system consisting of $A \to i_r A$ is transformed into an inverse system of the morphisms $z^r \leftarrow z$ on the set of reciprocal numbers. To take limits of invariants of $G_{n,c}$, we consider the limit of the latter inverse system, which is exactly our parameter set P introduced in Section 1.2.

Recall that P is the set of all sequences $\alpha = (\ldots, \alpha_2, \alpha_1)$ of reciprocal numbers α_c such that $(\alpha_{rc})^r = \alpha_c$ for any r and c. Sometimes we denote $\alpha = (\alpha_c)$. Let P_0 be the subset of P consisting of all $\alpha = (\alpha_c)$ such that $|\alpha_1| = 1$. Note that this implies $|\alpha_c| = 1$ for all c.

For $\mathcal{A} \in \mathcal{G}_n$, choose $[A] \in G_{n,c}$ which represents $\mathcal{A}$, i.e., $\mathcal{A}$ is the image of $[A]$ under the homomorphism $\phi_c \colon G_{n,c} \to \mathcal{G}_n$. We define the signature and rank invariants of $\mathcal{A}$ by

$$s(\mathcal{A}) = (s_{\alpha_c}[A])_{\alpha \in P_0} \quad \in \mathbb{Z}^{P_0},$$
$$e(\mathcal{A}) = (e_{\alpha_c}[A])_{\alpha \in P} \quad \in (\mathbb{Z}/2)^P.$$

To define our discriminant invariant, first we construct its codomain as follows. For $c \mid d$ and $\alpha = (\alpha_c) \in P$, $\mathbb{Q}(\alpha_c + \alpha_c^{-1})^\times$ is a subgroup of $\mathbb{Q}(\alpha_d + \alpha_d^{-1})^\times$. Taking the product of induced homomorphisms over all $\alpha \in P$, we obtain

$$\prod_{\alpha \in P} \frac{\mathbb{Q}(\alpha_c + \alpha_c^{-1})^\times}{N_{\alpha_c}^\times} \longrightarrow \prod_{\alpha \in P} \frac{\mathbb{Q}(\alpha_d + \alpha_d^{-1})^\times}{N_{\alpha_d}^\times}.$$

We consider the limit

$$\varinjlim_c \prod_{\alpha \in P} \frac{\mathbb{Q}(\alpha_c + \alpha_c^{-1})^\times}{N_{\alpha_c}^\times}$$

of the direct system consisting of the above homomorphisms. For $\mathcal{A} \in \mathcal{G}_n$ represented by $[A] \in G_{n,c}$ as before, we denote by $d(\mathcal{A})$ the element in the limit represented by

$$(d_{\alpha_c}[A])_{\alpha \in P} \in \prod_{\alpha \in P} \frac{\mathbb{Q}(\alpha_c + \alpha_c^{-1})^\times}{N_{\alpha_c}^\times}.$$

THEOREM 3.11. $s(\mathcal{A})$, $e(\mathcal{A})$, and $d(\mathcal{A})$ are well-defined invariants of $\mathcal{A} \in \mathcal{G}$.

PROOF. Suppose $[A] \in G_{n,c}$ and $[B] \in G_{n,d}$ are sent to the same element $\mathcal{A} \in \mathcal{G}_n$ by ϕ_c and ϕ_d, respectively. Then $i_{rd}[A] = i_{rc}[B]$ for some r. For $\alpha \in P$, we have

$$d_{\alpha_c}[A] = d_{(\alpha_{rcd})^{rd}}[A] \equiv d_{\alpha_{rcd}} i_{rd}[A]$$
$$= d_{\alpha_{rcd}} i_{rc}[B] \equiv d_{(\alpha_{rcd})^{rc}}[B] = d_{\alpha_d}[B] \mod N_{\alpha_{rcd}}^\times$$

by the reparametrization formula (Lemma 3.8). This shows that $(d_{\alpha_c}[A])_{\alpha \in P}$ and $(d_{\alpha_d}[B])_{\alpha \in P}$ give rise to the same element in the limit. Similar arguments work for $s(\mathcal{A})$ and $e(\mathcal{A})$. $\square$

As an immediate consequence of Proposition 3.6, we have the following additivity of our invariants. To state the additivity of d, we use the following notation: for an element $x = (x_\alpha)_{\alpha \in P} \in (\mathbb{Z}/2)^P$, let denote the element

$((-1)^{x_\alpha})_{\alpha \in P} \in \prod_{\alpha \in P}\{\pm 1\}$ by $(-1)^x$. Note that the multiplicative group $\prod_{\alpha \in P}\{\pm 1\}$ acts on

$$\prod_{\alpha \in P} \frac{\mathbb{Q}(\alpha_c + \alpha_c^{-1})^\times}{N_{\alpha_c}^\times}$$

by coordinatewise multiplication. It gives rise to an action of $\prod_{\alpha \in P}\{\pm 1\}$ on

$$\varinjlim_c \prod_{\alpha \in P} \frac{\mathbb{Q}(\alpha_c + \alpha_c^{-1})^\times}{N_{\alpha_c}^\times}.$$

PROPOSITION 3.12.
(1) $s(\mathcal{A} + \mathcal{B}) = s(\mathcal{A}) + s(\mathcal{B})$.
(2) $e(\mathcal{A} + \mathcal{B}) = e(\mathcal{A}) + e(\mathcal{B})$.
(3) $d(\mathcal{A} + \mathcal{B}) = (-1)^{e(\mathcal{A})e(\mathcal{B})} d(\mathcal{A}) d(\mathcal{B})$.

The remaining part of this section is devoted to the proof of the completeness of our invariants:

THEOREM 3.13. *An element $\mathcal{A} \in \mathcal{G}_n$ is trivial if and only if the invariants $s(\mathcal{A})$, $e(\mathcal{A})$, and $d(\mathcal{A})$ are trivial.*

The following observations are useful in proving Theorem 3.13.

LEMMA 3.14.
(1) *If $p(t)$ is non-reciprocal and irreducible, then $p(t^r)$ never has a reciprocal irreducible factor for all r.*
(2) *If w is a zero of an irreducible polynomial $p(t)$ and $q(t)$ is an irreducible factor of $p(t^r)$, then there is a zero z of $q(t)$ such that $z^r = w$.*

PROOF. (1) If an irreducible factor $q(t)$ of $p(t^r)$ is reciprocal, then $q(z) = 0 = q(z^{-1})$ for some z, and hence $p(z^r) = 0 = p(z^{-r})$. Thus $p(t)$ is reciprocal.

(2) Choose any zero z' of $q(t)$. Then $w' = (z')^r$ is a zero of $p(t)$ and so there is a Galois automorphism h on the algebraic closure of the base field such that $h(w') = w$. Let $z = h(z')$. Then z is a zero of $q(t)$ and $z^r = h(z')^r = h(w') = w$ as desired. $\square$

LEMMA 3.15. *Suppose $\lambda(t)$ is an irreducible polynomial and c is a positive integer. Then $\lambda(t^r)$ has a reciprocal irreducible factor for all positive integer r if and only if there exists $\alpha \in P$ such that $\lambda(\alpha_c) = 0$.*

PROOF. If $\lambda(\alpha_c) = 0$ for some α, then α_{rc} is a root of $\lambda(t^r)$. The irreducible polynomial of α_{rc} is reciprocal and divides $\lambda(t^r)$.

For the converse, we may assume $c = 1$ by coordinate shifting. Indeed, for any $\alpha = (\alpha_i)$ in P, $\alpha' = ((\alpha_i)^c)$ is also in P and its c-th coordinate is $(\alpha_c)^c = \alpha_1$.

First we choose a sequence $1 = n_1, n_2, n_3, \ldots$ of positive integers such that n_{i+1} is a multiple of n_i and every integer r divides some n_i. For example, we may enumerate all primes as $p_1, p_2, \ldots$ and put $n_{i+1} = p_1^i p_2^i \cdots p_i^i$.

We will construct a sequence of reciprocal irreducible polynomials $\lambda(t) = \lambda_1(t), \lambda_2(t), \ldots$ such that

(1) $\lambda_i(t^r)$ has a reciprocal irreducible factor for all r, and
(2) $\lambda_{i+1}(t)$ divides $\lambda_i(t^{n_{i+1}/n_i})$ for all $i \geq 1$

by an induction. Assume that $\lambda_i(t)$ has been chosen. Consider the irreducible factorization

$$\lambda_i(t^{n_{i+1}/n_i}) = \mu_1(t) \cdots \mu_k(t).$$

We claim that for at least one factor, say $\mu_1(t)$, $\mu_1(t^r)$ has a reciprocal irreducible factor for all r. Then we can put $\lambda_{i+1}(t) = \mu_1(t)$. Suppose the claim is not true. Then we can choose r_i such that all irreducible factors of $\mu_i(t^{r_i})$ are non-reciprocal. By Lemma 3.14 (1), for any common multiple r of the r_i,

$$\lambda_i(t^{rn_{i+1}/n_i}) = \mu_1(t^r) \cdots \mu_k(t^r)$$

has no reciprocal irreducible factor. It contradicts the induction hypothesis (1).

Now we will choose a certain zero α_{n_i} of $\lambda_i(t)$ inductively. Let α_{n_1} be any zero of $\lambda_1(t)$. Suppose α_{n_i} has been chosen. Since $\lambda_{i+1}(t)$ divides $\lambda_i(t^{n_{i+1}/n_i})$, we can choose a zero $\alpha_{n_{i+1}}$ of $\lambda_{i+1}(t)$ satisfying $(\alpha_{n_{i+1}})^{n_{i+1}/n_i} = \alpha_{n_i}$ by appealing to Lemma 3.14 (2). We note that the chosen numbers satisfy $(\alpha_{n_i})^{n_i/n_j} = \alpha_{n_j}$ for any $i > j$.

For any positive integer c, choose n_i divided by c and let $\alpha_c = (\alpha_{n_i})^{n_i/c}$. α_c is well-defined, independent of the choice of n_i; for, if c divides both n_i and n_j where $i > j$, then

$$(\alpha_{n_j})^{n_j/c} = ((\alpha_{n_i})^{n_i/n_j})^{n_j/c} = (\alpha_{n_i})^{n_i/c}.$$

Since α_{n_i} is reciprocal, so is α_c. Moreover, for any r and c, there is some n_i divided by rc, and

$$(\alpha_{rc})^r = ((\alpha_{n_i})^{n_i/rc})^r = (\alpha_{n_i})^{n_i/c} = \alpha_c.$$

Therefore $\alpha = (\alpha_c)$ is an element of P such that $\lambda(\alpha_1) = 0$. □

PROOF OF THEOREM 3.13. Suppose that $[A] \in G_{n,c}$ represents $\mathcal{A} \in \mathcal{G}_n$ and $s(\mathcal{A})$, $e(\mathcal{A})$, and $d(\mathcal{A})$ vanish. By replacing $[A]$ with $[i_2 A] \in G_{n,2c}$, we may assume that the (± 1)-primary parts of $[A]$ are trivial, since

$$\Delta_{i_2 A}(\pm 1) = \Delta_A((\pm 1)^2) = \Delta_A(1) \neq 0.$$

By definitions, $s_{\alpha_c}(A) = 0$ for $\alpha \in P_0$ and $e_{\alpha_c}(A) = 0$ for $\alpha \in P$. For d, there exists r such that $d_{\alpha_c}(A) \in N_{\alpha_{rc}}^\times$ for all $\alpha \in P$. By replacing $[A] \in G_{n,c}$ by $[i_r A] \in G_{n,rc}$, we may assume that $d_{\alpha_c}(A) \in N_{\alpha_c}^\times$ for any $\alpha \in P$ by the reparametrization formula (Lemma 3.8).

Without any loss of generality, we may assume that A has only one nontrivial primary part, say the z-primary part for some reciprocal $z \neq \pm 1$. Let $\lambda(t)$ be the irreducible polynomial of z. If $\lambda(t^r)$ has no symmetric factor for some r, then $[i_r A] = 0$ in G_n by Proposition 3.6 (1), and thus $\mathcal{A} = 0$

in $\mathcal{G}_n$. Unless, by Lemma 3.15, there is $\alpha \in P$ such that α_c is a zero of $\lambda(t)$. Thus by the above paragraph, $s_z[A]$ (when $|z| = 1$), $e_z[A]$, and $d_z[A]$ vanish. Since $z \neq \pm 1$, it follows that $[A] = 0$. It completes the proof. □

THEOREM 3.16. *Every element in $\mathcal{G}_n$ has order 1, 2, 4, or ∞.*

PROOF. Let $\mathcal{A}$ be an element in $\mathcal{G}_n$. By Proposition 3.12 (2) and (3), $e(4\mathcal{A})$ and $d(4\mathcal{A})$ are always trivial. If $s(\mathcal{A})$ is nontrivial, then $\mathcal{A}$ has infinite order by the additivity of s. Suppose that $s(\mathcal{A})$ is trivial. Then, $s(4\mathcal{A})$ is trivial by Proposition 3.12 (1). From this it follows that $4\mathcal{A} = 0$ in $\mathcal{G}_n$, by Theorem 3.13. □

In future sections, we will frequently use the following "coordinates" of our invariants. For $\mathcal{A} \in \mathcal{G}_n$ and $\alpha \in P_0$, we denote $s_\alpha(\mathcal{A}) \in \mathbb{Z}$ be the α-th coordinate of $s(\mathcal{A}) \in \mathbb{Z}^{P_0}$. For $\alpha \in P$, $e_\alpha(\mathcal{A}) \in \mathbb{Z}/2$ is defined similarly. We denote by $d_\alpha(\mathcal{A})$ the image of $d(A)$ under the canonical map

$$\varinjlim_c \prod_{\alpha \in P} \frac{\mathbb{Q}(\alpha_c + \alpha_c^{-1})^\times}{N_{\alpha_c}^\times} \longrightarrow \varinjlim_c \frac{\mathbb{Q}(\alpha_c + \alpha_c^{-1})^\times}{N_{\alpha_c}^\times}.$$

If $\mathcal{A}$ is the image of $[A] \in G_{n,c}$, then the coordinates can be described in terms of invariants of $[A]$: $s_\alpha(\mathcal{A}) = s_{\alpha_c}[A]$ and $e_\alpha(\mathcal{A}) = e_{\alpha_c}[A]$. $d_\alpha(A)$ is the image of $d_{\alpha_c}[A]$ under

$$\frac{\mathbb{Q}(\alpha_c + \alpha_c^{-1})^\times}{N_{\alpha_c}^\times} \longrightarrow \varinjlim_c \frac{\mathbb{Q}(\alpha_c + \alpha_c^{-1})^\times}{N_{\alpha_c}^\times}.$$

s_α, e_α, and d_α also have additivity properties similar to Proposition 3.12. For d_α, the additivity can be expressed as

$$d_\alpha(\mathcal{A} + \mathcal{B}) = (-1)^{e_\alpha(\mathcal{A})e_\alpha(\mathcal{B})} d_\alpha(\mathcal{A}) d_\alpha(\mathcal{B}).$$

3.3. Computation of $e(\mathcal{A})$

In the remaining part of this chapter we give a full calculation of the algebraic structure of $\mathcal{G}_n$ using our invariants discussed in the previous section. We remark that the torsion-free part of $\mathcal{G}_n$ is already well understood: in [12, 7] it was shown that there are infinitely many independent elements of infinite order in $\mathcal{G}_n$.

In this section we focus on the computation of order two elements in $\mathcal{G}_n$. Although there are 2 and 4-torsion elements in $G_{n,c}$ (e.g., the argument of Levine's work [32] on the structure of $G_n^{\mathbb{Z}}$ can be applied), it has been still unknown whether there are nontrivial torsion elements in $\mathcal{G}_n$ since $G_{n,c} \to \mathcal{G}_n$ may kill torsion elements.

EXAMPLE 3.17 (Kawauchi [22]). Consider a Seifert matrix

$$A = \begin{bmatrix} 1 & 1 \\ 0 & -1 \end{bmatrix}$$

of the figure eight knot. It is well known $[A] \in \mathcal{G}_n = \mathcal{G}_{n,c}$ has order 2 (when $n \equiv 1 \mod 4$). However, since the irreducible factors of the Alexander polynomial
$$\Delta_{i_2 A}(t) = \Delta_A(t^2) = (t^2 - t - 1)(t^2 + t - 1)$$
are all non-reciprocal, $[i_2 A] = 0$. Therefore the image of $[A]$ in $\mathcal{G}_n$ is trivial.

In the above example, the key property is that $\Delta_A(t^r)$ is factored into non-reciprocal factors for some r. For any such A, the image of $[A]$ in $\mathcal{G}_n$ is trivial. Appealing to Lemma 3.15, this property may be rephrased as follows: for any $\alpha = (\alpha_i) \in P$ and $c > 0$, α_c is not a zero of $\Delta_A(t)$. Thus, in order to obtain a nontrivial element in $\mathcal{G}_n$, we need to construct A such that $\Delta_A(\alpha_c) = 0$ for some $\alpha = (\alpha_i) \in P$. Of course when $\Delta_A(t)$ has a zero of unit complex length, we can easily find such an $\alpha \in P_0$. However, in this case, α may have nontrivial contribution to the signature invariant so that the order is not finite.

The first step of our construction of nontrivial torsion elements in $\mathcal{G}_n$ is to find elements in $P - P_0$ which automatically have no contribution to the signature.

PROPOSITION 3.18. *Suppose* $\lambda(t) = at^2 - (2a + p)t + a$, *where* a *is a prime and* p *an integer such that* $p \not\equiv 0 \mod a$ *and* $p \not\equiv -2a \pm 1 \mod a^2$. *Then* $\lambda(t^r)$ *is irreducible for any positive integer* r.

PROOF. Our proof consists of elementary arguments. Suppose
$$\lambda(t^r) = at^{2r} - (2a + p)t^r + a$$
$$= (b_k t^k + \cdots + b_1 t + b_0)(c_l t^l + \cdots c_1 t + c_0)$$
where b_i and c_j are integers, $k + l = 2r$, and $k, l < 2r$.

Since $b_0 c_0 = a$ is a prime, we may assume that $b_0 = 1$ and $c_0 = a$. By looking at the coefficients of $t^1, t^2, \ldots$, we have
$$0 = b_0 c_i + b_1 c_{i-1} + \cdots + b_i c_0$$
for $1 \leq i \leq r - 1$, and hence inductively
$$0 \equiv c_0 \equiv \cdots \equiv c_{r-1} \mod a.$$
Similarly $c_r \equiv -p \not\equiv 0 \mod a$.

Computing the coefficients of $t^{2r}, t^{2r-1}, \ldots$ from the higher degree terms of the two factors, we have
$$0 = b_k c_{l-i} + b_{k-1} c_{l-i+1} + \cdots + b_{k-i} c_l$$
for $1 \leq i \leq r - 1$ and
$$-(2a + p) = b_k c_{l-r} + b_{k-1} c_{l-r+1} + \cdots + b_{k-r} c_l.$$
Since $a = b_k c_l$ is a prime, we have two cases.

Case 1: $b_k = \pm 1$ and $c_l = \pm a$. Then
$$c_{l-1} \equiv \cdots \equiv c_{l-r+1} \equiv 0,$$
$$c_{l-r} \equiv \pm p \not\equiv 0 \mod a.$$

Thus $r \leq l - r$. It contradicts $l < 2r$.

Case 2: $b_k = \pm a$ and $c_l = \pm 1$. Then similarly
$$b_k \equiv \cdots \equiv b_{k-r+1} \equiv 0,$$
$$b_{k-r} \not\equiv 0 \mod a.$$

Thus $k - r \geq 0$. Since $c_l \not\equiv 0$, $l \geq r$. Hence $k = l = r$. Now looking at the t^r term, we have
$$-(2a + p) \equiv b_r c_0 + b_{r-1} c_1 + \cdots + b_0 c_r \equiv b_0 c_r = \pm 1 \mod a^2.$$

It contradicts the hypothesis. $\square$

We have the following consequence of Proposition 3.18 and Lemma 3.15.

COROLLARY 3.19. *Suppose $\lambda(t)$ is as in Proposition 3.18, z is a zero of $\lambda(t)$, and c is a positive integer. Then there is $\alpha = (\alpha_i) \in P$ such that $\alpha_c = z$.*

REMARK 3.20.
(1) The polynomial in Proposition 3.18 has two different real zeros which are not ± 1 if $p(4a + p) > 0$. In particular, the element α in Corollary 3.19 is not in P_0.
(2) There are infinitely many pairs (a, p) satisfying the assumption of Proposition 3.18. For example, if $a > 3$ is a prime and $0 < p < a$, then (a, p) satisfies the assumption.

REMARK 3.21. If $\lambda(t) = at^2 - (2a + p)t + a$ is as in Proposition 3.18, then the conclusion of Proposition 3.18 also holds for $\lambda(-t)$. For, it is easily seen that (a, p) satisfies the assumptions of Proposition 3.18 if and only if so does $(a, -4a - p)$, and thus we can apply Proposition 3.18 for
$$\lambda(-t) = at^2 + (2a + p)t + a = at^2 - (2a + (-4a - p))t + a.$$

In order to construct Seifert matrices whose Alexander polynomials are as in Proposition 3.18, we appeal to the following general characterization and realization theorem for Alexander polynomials. Recall from Section 2.2 that a rational Seifert matrix is defined to be a square matrix A such that $P(A - \epsilon A^T)P^T$ is integral, even, and unimodular over $\mathbb{Z}$ for some rational square matrix P. Note that a rational Seifert matrix is always of even dimension.

THEOREM 3.22. *A polynomial $\Delta(t)$ is the Alexander polynomial of some $2g \times 2g$ rational Seifert matrix if and only if*
(1) $\Delta(t) = \Delta(t^{-1})t^{2g}$,
(2) $\Delta(\epsilon)$ *is a square in* $\mathbb{Q}$, *and*
(3) $\epsilon^g \Delta(1)$ *is a nonzero square in* $\mathbb{Q}$.

REMARK 3.23. There is a well-known characterization of the Alexander polynomial of an integral Seifert matrix A (i.e., $A - \epsilon A^T$ is unimodular) by Levine [33]. Our characterization gives a larger class of polynomials than integral Alexander polynomials. For example, there are integral polynomials $\Delta(t)$ which can be realized as Alexander polynomials of rational Seifert matrices but $\Delta(1) \neq \pm 1$. No integral Seifert matrix has such an Alexander polynomial.

PROOF. Our argument is similar to [33]. Suppose that $\Delta(t) = \det(tA - \epsilon A^T)$ for some $2g \times 2g$ rational Seifert matrix A. (1) is immediate. Since $\epsilon A - \epsilon A^T$ is skew-symmetric, (2) follows. For (3), $\Delta(1) \neq 0$ since $A - \epsilon A^T$ is nonsingular. If $\epsilon = 1$, (2) implies (3). If $\epsilon = -1$, the signature of $U = P(A + A^T)P^T$ is known to be divisible by 8 and so

$$\Delta(1) = \det(A + A^T) \equiv \det(U) = (-1)^g$$

modulo squares.

For the converse, we will use an induction on g to show that $\Delta(t) = \det(tA - \epsilon A^T)$ for some rational Seifert matrix A satisfying the following auxiliary condition: the $(1,1)$-cofactor of $tA - \epsilon A$ is

$$(-\epsilon)^{g-1}(t-1)^{2g-2}(t-\epsilon).$$

For $g = 1$, $\Delta(t)$ is of the form $at^2 + bt + a$. If $\epsilon = 1$, $\Delta(1) = u^2$ implies that $b = u^2 - 2a$, where $u \in \mathbb{Q}^\times$. Then it can be verified that

$$A = \begin{bmatrix} a & u \\ 0 & 1 \end{bmatrix}$$

satisfies all the desired properties. If $\epsilon = -1$, $\Delta(-1) = u^2$ and $\Delta(1) = -v^2$ for some $u \in \mathbb{Q}$, $v \in \mathbb{Q}^\times$ and hence $a = (u^2 - v^2)/4$, $b = -(u^2 + v^2)/2$. Then we can take

$$A = \begin{bmatrix} \frac{u^2-v^2}{4} & u \\ 0 & 1 \end{bmatrix}.$$

In this case, for

$$P = \begin{bmatrix} \frac{1}{v} & \frac{-u+v}{2v} \\ -\frac{1}{v} & \frac{u+v}{2v} \end{bmatrix},$$

$P(A + A^T)P^T$ is an even unimodular integral matrix.

Now suppose $g > 1$. Given $\Delta(t)$ satisfying the above (1)–(3), let $a = -(-\epsilon)^{g-1}\Delta(0)$, and choose $\Delta_0(t)$ such that

$$\Delta(t) = -a(-\epsilon)^{g-1}(t-1)^{2g-2}(t-\epsilon)^2 + \epsilon t \Delta_0(t).$$

Then it can be checked that $\Delta_0(t)$ satisfies (1)–(3) where $g - 1$ plays the role of g. By our induction hypothesis, $\Delta_0(t) = \det(tA_0 - \epsilon A_0^T)$ for some $(2g-2) \times (2g-2)$ rational Seifert matrix A_0 satisfying the auxiliary condition. Let

$$A = \left[\begin{array}{cc|c} 0 & 1 & a \\ 0 & 0 & 1 \\ \hline \epsilon a & \epsilon & \\ & & A_0 \end{array} \right].$$

Then we can check that $\det(tA - \epsilon A) = \Delta(t)$ and A satisfies all the desired properties including our auxiliary condition. It completes the proof. □

Using the above results, we can construct examples with nontrivial $e(\mathcal{A})$. Let
$$\lambda(t) = -\frac{a}{p}t^2 + \left(\frac{2a}{p} + 1\right)t - \frac{a}{p}$$
where a and p are nonzero integers. Then from Theorem 3.22 it follows that
$$\Delta(t) = \begin{cases} \lambda(t) & \text{for } \epsilon = 1 \\ (t+1)^2\lambda(t) & \text{for } \epsilon = -1 \end{cases}$$
is always the Alexander polynomial of a rational Seifert matrix.

THEOREM 3.24. *Suppose a and p are positive integers satisfying the hypothesis of Proposition 3.18 and A is a rational Seifert matrix whose Alexander polynomial is $\Delta(t)$ given above. Then the image $\mathcal{A}$ of $[A]$ under $\phi_c\colon \mathcal{G}_n = \mathcal{G}_{n,c} \to \mathcal{G}_n$ has order 2 or 4 for any c.*

PROOF. First we claim that $s(\mathcal{A}) = 0$. For $\epsilon = 1$, it follows from the fact that $\Delta(t)$ has no zero of unit length. For $\epsilon = -1$, although $z = -1$ is a zero of $\Delta(t)$, it has no contribution to the signature $s(\mathcal{A})$. (In fact, the (-1)-primary part is a skew-symmetric form over $\mathbb{Q}(-1) = \mathbb{Q}$ and hence automatically Witt trivial.)

From the claim, $\mathcal{A}$ has finite order in $\mathcal{G}_n$. Thus it suffices to show that $\mathcal{A}$ is nontrivial. Let z be a zero of $\lambda(t)$. Then by Corollary 3.19, there is an element $\alpha = (\alpha_c) \in P$ such that $\alpha_c = z$. By Proposition 3.6, the α-th coordinate $e_\alpha(\mathcal{A}) = e_{\alpha_c}[A] \in \mathbb{Z}/2$ of $e(\mathcal{A})$ is the exponent of the irreducible polynomial $\lambda(t)$ of z in $\Delta(t)$, which is equal to 1. Thus $e(\mathcal{A})$ is nontrivial. □

REMARK 3.25. In some cases we can explicitly determine the order of $\mathcal{A}$ constructed above without computing $d(\mathcal{A})$. Suppose $\epsilon = 1$, a is an odd prime, and $p = 1$. Then in the above proof we can choose A as an integral Seifert matrix of an n-knot in S^{n+2} with Alexander polynomial $\Delta(t)$; for, it is easily checked that our polynomial $\Delta(t)$ has integral coefficients and satisfies the conditions of Levine's realization theorem [**33**, Proposition 1]. It is known that if every prime $\equiv -1 \mod 4$ has an even exponent in the factorization of $4a + 1$, then $[A]$ has order 2 in $\mathcal{G}_n$ (e.g. see [**32**, Corollary 23 (c)]). Thus, for such a, $\mathcal{A}$ has order 2 in $\mathcal{G}_n$.

In particular, if both a and $4a+1$ are primes, then $\mathcal{A}$ has order 2 in $\mathcal{G}_n$. This relates the structure of $\mathcal{G}_n$ with a well-known open problem in number theory: are there infinitely many pairs of primes of the form $(a, 4a+1)$? If the answer is affirmative, then Theorem 3.24 can be used, without computing $d(\mathcal{A})$, to produce infinitely many order 2 elements that generate a $(\mathbb{Z}/2)^\infty$ summand of $\mathcal{G}_n$.

At present, the author does not know any method to construct $(\mathbb{Z}/2)^\infty$ and $(\mathbb{Z}/4)^\infty$ summands of $\mathcal{G}_n$ without computing $d(\mathcal{A})$. In the next two sections we compute $d(\mathcal{A})$ explicitly.

3.4. Artin reciprocity and norm residue symbols

The most crucial difficulty in the computation of $d(\mathcal{A})$ is to detect nontrivial elements in the limit

$$\varinjlim_{c} \prod_{\alpha \in P} \frac{\mathbb{Q}(\alpha_c + \alpha_c^{-1})^\times}{N_{\alpha_c}^\times}$$

where the value of $d(\mathcal{A})$ lives. To study the limit, first we consider an easier problem whether an element in $\mathbb{Q}(z+z^{-1})^\times$ is contained in

$$N_z^\times = \{w\bar{w} \mid w \in \mathbb{Q}(z)^\times\}.$$

The main tool we will use is the Artin reciprocity, which is one of the central machinery in algebraic number theory. In this section, for readers not familiar with this, we give a quick review of necessary results from algebraic number theory, which can be used as a reference for later sections. We claim no originality on the materials discussed in this section. There are several good general references on algebraic number theory, e.g., [**4, 42, 29**].

In the next section, we will investigate the *limiting* behaviour of the Artin reciprocity, as an interesting application of these number theoretic tools which is related with the structure of limits of Seifert matrices.

Let L be an abelian extension of a number field K and let $N_K^L \colon L^\times \to K^\times$ be the norm. Of course the main example we keep in mind is $L = \mathbb{Q}(z)$ and $K = \mathbb{Q}(z + z^{-1})$ where z is a reciprocal number. In this case the (multiplicative) subgroup $N_z^\times$ in $\mathbb{Q}(z+z^{-1})^\times$ can be identified with the group of nonzero norms for L over K, i.e., the image of N_K^L. Motivated from this, we consider the problem of detecting nontrivial elements of $K^\times/N_K^L(L^\times)$. By the Hasse principle, the global problem can be reduced into a local problem. For a valuation v of K, we denote the completion of K with respect to v by K_v and the completion of L with respect to an extension of v by L^v.

THEOREM 3.26 (Hasse principle). *An element x in $K^\times$ is a norm for L over K if and only if x is a norm for K_v over L^v for every valuation v on K.*

In the local case, the local Artin reciprocity relates norms with an associated Galois group.

THEOREM 3.27 (Local Artin reciprocity). *There is a surjection*

$$\varphi_v \colon K_v^\times \longrightarrow \operatorname{Gal}(L^v/K_v)$$

whose kernel is the set of nonzero norms of L^v over K_v.

The above homomorphism φ_v is called the *local Artin map*. From this the group $K_v^\times/N_{K_v}^{L^v}(L^{v\times})$ can be identified with the Galois group $\operatorname{Gal}(L^v/K_v)$, and hence our problem can be solved by computing φ_v and the structure of $\operatorname{Gal}(L^v/K_v)$.

3.4. ARTIN RECIPROCITY AND NORM RESIDUE SYMBOLS

Furthermore, for the case of a Kummer extension, this association can be described in terms of the norm residue symbols (or Hilbert symbols). Although it can be done for any Kummer extension, we will focus on quadratic extensions $L = K(\sqrt{a})$ ($a \in K$), which will be sufficient for our purpose. In this case $L^v = K(\sqrt{a})^v$ is isomorphic to $K_v(\sqrt{a})$ and $\pm\sqrt{a}$ are all the conjugates of $\sqrt{a}$. Thus a Galois automorphism of L^v over K_v is either the identity or $\sqrt{a} \to -\sqrt{a}$.

DEFINITION 3.28. For $a, b \in K^\times$, the *(quadratic) norm residue symbol* $(a, b)_v$ is defined by the equation

$$\varphi_v(b)(\sqrt{a}) = (a, b)_v \sqrt{a}$$

where $\varphi_v \colon K_v^\times \to \operatorname{Gal}(K_v(\sqrt{a})/K_v)$ is the local Artin map discussed above.

Obviously $(a, b)_v = \pm 1$. We summarize basic properties of the norm residue symbol:

PROPOSITION 3.29 ([4, 42]).
(1) $(\ ,\ )_v \colon K^\times \times K^\times \to \{\pm 1\}$ *is symmetric and bilinear.*
(2) $(a, b)_v = 1$ *if and only if b is a norm for $K_v(\sqrt{a})$ over K_v.*
(3) *For any $a, b \in K^\times$, $(a, b)_v = 1$ for all but finitely many v. Furthermore $\prod (a, b)_v = 1$ where v runs over all valuations on K.*

A consequence of Proposition 3.29 and the Hasse principle is that b is a norm for $K(\sqrt{a})$ over K if and only if $(a, b)_v = 1$ for every v.

For computation, the following results are useful:

PROPOSITION 3.30 ([4, 42]).
(1) *If $K_v = \mathbb{C}$, then $(a, b)_v = 1$ for any a, b.*
(2) *If $K_v = \mathbb{R}$, then $(a, b)_v = 1$ if and only if $a > 0$ or $b > 0$.*
(3) *If v is a non-archimedean valuation associated to a prime $\mathfrak{p}$ of K over an odd prime $p \in \mathbb{Z}$, then*

$$(a, b)_v = \left((-1)^{v(a)v(b)} \frac{a^{v(b)}}{b^{v(a)}} \right)^{\frac{p^{f(\mathfrak{p},p)}-1}{2}}$$

where $f(\mathfrak{p}, p) = [\mathcal{O}_K/\mathfrak{p} : \mathbb{Z}/p]$ is the degree of the residue field extension $\mathcal{O}_K/\mathfrak{p}$ over $\mathbb{Z}/p$, and $\mathcal{O}_K$ is the ring of integers of K. (The right-hand side is viewed as a formula in $\mathcal{O}_K/\mathfrak{p}$.)
(4) *If $K = \mathbb{Q}$ and v is the 2-adic norm, then*

$$(a, b)_v = (-1)^{e(a')e(b') + v(a)w(b') + v(b)w(a')}$$

where $a = 2^{v(a)}a'$, $b = 2^{v(b)}b'$, and $e(x)$ and $w(x)$ denote the modulo 2 residue class of $(x-1)/2$ and $(x^2-1)/8$, respectively.

REMARK 3.31. In the case of a general number field K other than $\mathbb{Q}$, the computation of $(a, b)_v$ for a valuation v associated to a prime $\mathfrak{p}$ over 2 is more complicated. We will not use this.

In our computation of the norm residue symbols using Proposition 3.30, we will use the following results on splittings of primes. Suppose A is a Dedekind domain with quotient field K, L is a finite extension over K of degree n, and B is the integral closure of A in L. For a prime $\mathfrak{p}$ in A, let

$$\mathfrak{p}B = \prod_{\mathfrak{q}|\mathfrak{p}} \mathfrak{q}^{e(\mathfrak{q},\mathfrak{p})}$$

be the splitting in B where $\mathfrak{q}$ runs over the primes of B over $\mathfrak{p}$. Let $f(\mathfrak{q},\mathfrak{p})$ be the degree of $B/\mathfrak{q}$ over $A/\mathfrak{p}$. Then we have

LEMMA 3.32 ([**4, 42, 29**]).
(1) $n = \sum_{\mathfrak{q}|\mathfrak{p}} e(\mathfrak{q},\mathfrak{p}) f(\mathfrak{q},\mathfrak{p})$.
(2) $e(\mathfrak{r},\mathfrak{p}) = e(\mathfrak{r},\mathfrak{q}) e(\mathfrak{q},\mathfrak{p})$ and $f(\mathfrak{r},\mathfrak{p}) = f(\mathfrak{r},\mathfrak{q}) f(\mathfrak{q},\mathfrak{p})$ if $\mathfrak{r}$ is over $\mathfrak{q}$ and $\mathfrak{q}$ is over $\mathfrak{p}$.

For quadratic extensions, the splitting of a prime is well understood:

LEMMA 3.33 ([**42**]). *Suppose $L = K(\sqrt{\sigma})$ for some $\sigma \in A$ which is not square. If $B = A[\sqrt{\sigma}]$ is the integral closure of A in L and $\mathfrak{p}$ is a prime in A which does not contain 2, then $\mathfrak{p}B$ splits into primes in B as follows:*

$$\mathfrak{p}B = \begin{cases} (\mathfrak{p}, \sqrt{\sigma})^2 & \text{if } \sigma \equiv 0 \mod \mathfrak{p} \\ \mathfrak{p}B & \text{if } \sigma \not\equiv x^2 \mod \mathfrak{p} \text{ for any } x \in A \\ (\mathfrak{p}, x - \sqrt{\sigma})(\mathfrak{p}, x + \sqrt{\sigma}) & \text{if } 0 \not\equiv \sigma \equiv x^2 \mod \mathfrak{p} \text{ for some } x \in A \end{cases}$$

In the last case, $(\mathfrak{p}, x - \sqrt{\sigma})$ and $(\mathfrak{p}, x + \sqrt{\sigma})$ are different primes in B.

For the computation in later sections, we also need the following generalization of the last case of Lemma 3.33.

LEMMA 3.34. *Suppose A, K, σ, L, and $\mathfrak{p}$ are as in Lemma 3.33, and suppose B is the integral closure of A in L (without assuming $B = A[\sqrt{\sigma}]$). If $0 \not\equiv \sigma \equiv x^2 \mod \mathfrak{p}$ for some $x \in A$, then $\mathfrak{p}B$ splits into the product of two different primes $(\mathfrak{p}, x - \sqrt{\sigma})$ and $(\mathfrak{p}, x + \sqrt{\sigma})$.*

Since the author could not find a proof of Lemma 3.34 in the literature, he gives a proof for concreteness.

PROOF. By [**42**, Proposition III.12], the integral closure C of $A_\mathfrak{p}$ (localization of A away from $\mathfrak{p}$) in L is given by $C = A_\mathfrak{p}[\beta]$ for some $\beta \in C$. Since L is quadratic, $\beta^2 + a\beta = b$ for some $a, b \in A_\mathfrak{p}$. Since $1/2 \in A_\mathfrak{p}$, we may assume $a = 0$ by completing the square, i.e., $\beta = \sqrt{b}$. Since $\sqrt{\sigma} \in B \subset C = A_\mathfrak{p}[\sqrt{b}]$, we can write $\sqrt{\sigma} = u + v\sqrt{b}$ for some $u, v \in A_\mathfrak{p}$. From $\sigma = u^2 + v^2 b + 2uv\sqrt{b}$, it follows that $u = 0$ since $v = 0$ implies that σ is square. Writing $v = s/r$ where $s \in A$, $r \in A - \mathfrak{p}$, we have $r^2 \sigma = s^2 b$. Since r and σ are not in $\mathfrak{p}$, so are s and b, i.e., $v_\mathfrak{p}(v) = 0$. Therefore

$$b = v^{-2}\sigma \equiv (v^{-1}x)^2 \not\equiv 0 \mod \mathfrak{p}_\mathfrak{p}.$$

Now by Lemma 3.33, $\mathfrak{p}_\mathfrak{p}$ splits into two different primes in C, and thus so does $\mathfrak{p}$ in B.

On the other hand, from $2\sigma \notin \mathfrak{p}$, it follows that $\mathfrak{p} + (2\sigma) = (1)$ and $\mathfrak{p}^2 + 2\sigma\mathfrak{p} = \mathfrak{p}$ in A. Therefore

$$(\mathfrak{p}, x - \sqrt{\sigma}) \cdot (\mathfrak{p}, x + \sqrt{\sigma}) = (\mathfrak{p}^2, (x - \sqrt{\sigma})\mathfrak{p}, (x + \sqrt{\sigma})\mathfrak{p}, x^2 - \sigma)$$
$$\supset (\mathfrak{p}^2, 2\sqrt{\sigma}\mathfrak{p}) \supset (\mathfrak{p}^2, 2\sigma\mathfrak{p}) \supset \mathfrak{p}B$$

in B, that is, $(\mathfrak{p}, x - \sqrt{\sigma}) \cdot (\mathfrak{p}, x + \sqrt{\sigma})$ divides $\mathfrak{p}B$. This completes the proof. □

We finish this section with the following elementary lemma, which will be used in the next section for the computation of our discriminant invariant.

LEMMA 3.35. *Suppose K is a finite extension of a number field F.*
(1) *If $\mathfrak{p}$ is a prime of K which is over a prime $\mathfrak{q}$ of F, then*

$$f(\mathfrak{p}, \mathfrak{q}) v_\mathfrak{p}(-) = v_\mathfrak{q}(N_{F_{v_\mathfrak{q}}}^{K_{v_\mathfrak{p}}}(-))$$

where $v_\mathfrak{p}$ and $v_\mathfrak{q}$ are the valuations associated to the primes $\mathfrak{p}$ and $\mathfrak{q}$, respectively.
(2) *For any prime $\mathfrak{q}$ of F,*

$$\prod_{\mathfrak{p} | \mathfrak{q}} N_{F_{v_\mathfrak{q}}}^{K_{v_\mathfrak{p}}}(-) = N_F^K(-)$$

where $\mathfrak{p}$ runs over all primes of K which are over $\mathfrak{q}$.

For a proof, see [**4**] or [**42**].

3.5. Computation of $d(\mathcal{A})$

3.5.1. 2-torsion. Returning to the study of the torsion elements of $\mathcal{G}_n$, we will show that there are infinitely many order 2 elements. Consider a specialization of the polynomial $\Delta(t)$ used in Theorem 3.24 obtained by letting $p = 1$:

$$\Delta(t) = \begin{cases} \lambda(t) & \text{for } \epsilon = 1 \\ (t+1)^2 \lambda(t) & \text{for } \epsilon = -1 \end{cases}$$

where

$$\lambda(t) = -at^2 + (2a+1)t - a.$$

As in Section 3.3, it is the Alexander polynomial of a rational Seifert matrix A.

THEOREM 3.36. *If a is a prime such that $a \equiv 1 \mod 4$ and A is a rational Seifert matrix whose Alexander polynomial is $\Delta(t)$ given above, then the image $\mathcal{A}$ of $[A]$ under $G_n = G_{n,c} \to \mathcal{G}_n$ has order 2 for any c.*

PROOF. By Theorem 3.24, $\mathcal{A} \neq 0$. Thus it suffices to show that $\mathcal{A} + \mathcal{A} = 0$ in $\mathcal{G}_n$. The invariants s and e vanish for $\mathcal{A} + \mathcal{A}$ by additivity (Proposition 3.12). We will show that $i_2[A \oplus A]$ has vanishing discriminant invariants, i.e.,

$$d_z(i_2[A \oplus A]) \equiv 1 \mod N_z^\times$$

for all z. Then it follows that $d(\mathcal{A}+\mathcal{A})$ is trivial, and the proof is completed. (Indeed $i_2[A \oplus A] = 0$ is proved.)

Note that $i_2(A \oplus A)$ has Alexander polynomial $\Delta(t^2)^2$. Since $\lambda(t^2)$ is irreducible, the $\lambda(t^2)$-primary part is the only nontrivial primary part of $i_2(A \oplus A)$. (Note that for $\epsilon = -1$ the $(t+1)$-primary part of $i_2(A \oplus A)$ is automatically trivial.) Hence, in order to show the above claim, it suffices to consider a zero
$$z = \frac{\sqrt{4a+1}+1}{\sqrt{4a}}$$
of $\lambda(t^2)$. By Proposition 3.6,
$$d_z(i_2[A \oplus A]) \equiv -1 \mod N_z^\times.$$
So the question is whether -1 is a norm for the extension $L = \mathbb{Q}(z)$ over $K = \mathbb{Q}(z + z^{-1})$. Straightforward calculation shows that $K = \mathbb{Q}(\sqrt{a(4a+1)})$ and $L = K(z - z^{-1}) = K(\sqrt{a})$. Thus, by Section 3.4, we need to show $(-1, a)_v = 1$ for every valuation v on K. We consider the following three cases:

Case 1: Suppose v is archimedian, i.e., $K_v \cong \mathbb{R}$ or $\mathbb{C}$. Then $(-1, a)_v = 1$ by Proposition 3.30.

Case 2: Suppose v is induced by a prime $\mathfrak{p}$ of K which divides 2. If $\sqrt{a}$ is contained in K_v, then -1 is automatically a norm for $K_v(\sqrt{a})$ over K_v, and we are done. Assume not. We consider the following diagram of quadratic field extensions:

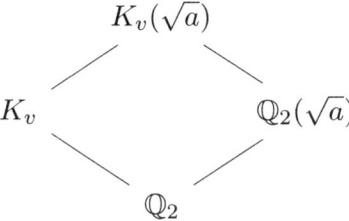

where $\mathbb{Q}_2$ is the 2-adic completion of $\mathbb{Q}$. Since the norm
$$N_{\mathbb{Q}_2}^{\mathbb{Q}_2(\sqrt{a})}: \mathbb{Q}_2(\sqrt{a})^\times \longrightarrow \mathbb{Q}_2^\times$$
is the restriction of
$$N_{K_v}^{K_v(\sqrt{a})}: K_v(\sqrt{a})^\times \longrightarrow K_v^\times,$$
it suffices to show that -1 is a norm for $\mathbb{Q}_2(\sqrt{a})$ over $\mathbb{Q}_2$. By Proposition 3.30,
$$(-1, a)_2 = (-1)^{e(-1)e(a)} = (-1)^{e(a)} = 1$$
since $a \equiv 1 \mod 4$.

Case 3: Suppose v is induced by a prime $\mathfrak{p}$ of K which divides an odd prime $p \in \mathbb{Z}$. Then by Proposition 3.30,
$$(-1, a)_v = (-1)^{v(a) \frac{p^{f(\mathfrak{p},p)}-1}{2}}.$$

We will show $v(a)$ is even. $\mathfrak{p}$ divides a if and only if p divides a. So if $p \neq a$, then $v(a) = 0$. Hence we may assume that $p = a$. Now by Lemma 3.33, $p = a$ splits into $\mathfrak{p}^2$ in $K = \mathbb{Q}(\sqrt{a(4a+1)})$ where $\mathfrak{p} = (a, \sqrt{a(4a+1)})$. (Note that $\mathbb{Q}$ and our K play the roles of K and L in Lemma 3.33, respectively.) Therefore $v(a) = 2$ and $(-1, a)_v = 1$ as desired. □

3.5.2. 4-torsion. Now we construct 4-torsion elements in $\mathcal{G}_n$. We again use the polynomials considered in Theorem 3.24: let

$$\Delta(t) = \begin{cases} \lambda(t) & \text{for } \epsilon = 1, \\ (t+1)^2 \lambda(t) & \text{for } \epsilon = -1, \end{cases}$$

where

$$\lambda(t) = -\frac{a}{p}t^2 + \left(\frac{2a}{p} + 1\right)t - \frac{a}{p}.$$

THEOREM 3.37. *Suppose that a and p are different primes such that $p \equiv -1 \mod 4$ and $p \not\equiv -(2a+1) \mod a^2$. If A is a rational Seifert matrix whose Alexander polynomial is $\Delta(t)$ given above, then the image $\mathcal{A}$ of $[A]$ under $G_n = G_{n,c} \to \mathcal{G}_n$ has order 4 for any c.*

PROOF. Because $\mathcal{A}$ is not of infinite order, it is of order 1, 2 or 4. Thus it suffices to show that $\mathcal{A} + \mathcal{A}$ is nontrivial. Let z be a zero of $\lambda(t)$. Choose $\alpha \in P$ such that $\alpha_c = z$ by appealing to Corollary 3.19. This gives us a choice of an r-th root of z for each r: we denote $z^{1/r} = \alpha_{cr}$. We will show that the "α-th coordinate" $d_\alpha(\mathcal{A} + \mathcal{A})$ of $d(\mathcal{A} + \mathcal{A})$ is nontrivial. By Proposition 3.12, we have

$$d_\alpha(\mathcal{A} + \mathcal{A}) = (-1)^{e_z[A]} = -1 \in \varinjlim_i \frac{\mathbb{Q}(\alpha_i + \alpha_i^{-1})^\times}{N_{\alpha_i}^\times}.$$

Thus, appealing to Section 3.4, it suffices to show that -1 is not a norm for $\mathbb{Q}(z^{1/r})$ over $\mathbb{Q}(z^{1/r} + z^{-1/r})$ for all r.

As a special case, suppose that $r = 2^k$ for some k. In this case we decompose the extension $\mathbb{Q}(z^{1/r} + z^{-1/r})$ over $\mathbb{Q}$ into a tower of quadratic extensions which are easier to understand. For notational convenience, let denote

$$K_i = \mathbb{Q}(z^{1/2^i} + z^{-1/2^i}),$$
$$L_i = \mathbb{Q}(z^{1/2^i}) = K_i(z^{1/2^i} - z^{-1/2^i}).$$

Let

$$m_i = \begin{cases} (2a+p)^2 & \text{for } i = 0, \\ a(2a + \sqrt{m_{i-1}}) & \text{for } i > 0. \end{cases}$$

Then it can be checked inductively that $z^{1/2^i} + z^{-1/2^i} = \sqrt{m_i}/a$ and so $K_i = \mathbb{Q}(\sqrt{m_i})$. Furthermore, since

$$(z^{1/2^i} - z^{-1/2^i})^2 = (z^{1/2^i} + z^{-1/2^i})^2 - 4 = m_i/a^2 - 4,$$

we have $L_i = K_i(\sqrt{\sigma_i})$ where $\sigma_i = m_i - 4a^2$.

We will construct a prime $\mathfrak{p}_i$ in the ring of integers $\mathcal{O}_{K_i}$ such that $m_i \equiv 4a^2 \mod \mathfrak{p}_i$, using an induction. Let $\mathfrak{p}_0 = (p)$. Then
$$m_0 = (2a+p)^2 \equiv 4a^2 \mod \mathfrak{p}_0$$
as desired. Suppose $\mathfrak{p}_i$ has been defined. Consider the splitting of $\mathfrak{p}_i$ in K_{i+1}: since $p \nmid 4a^2$, $m_i \equiv 4a^2 \not\equiv 0 \mod \mathfrak{p}_i$. Since $K_{i+1} = K_i(\sqrt{m_i})$,
$$\mathfrak{p}_i \mathcal{O}_{K_{i+1}} = (\mathfrak{p}_i, \sqrt{m_i} - 2a)(\mathfrak{p}_i, \sqrt{m_i} + 2a)$$
by Lemma 3.34. Let $\mathfrak{p}_{i+1} = (\mathfrak{p}_i, \sqrt{m_i} - 2a)$. Then
$$m_{i+1} = a(2a + \sqrt{m_i}) \equiv 4a^2 \mod \mathfrak{p}_{i+1}$$
as desired.

Let v_i be the valuation on K_i associated to the prime $\mathfrak{p}_i$. We will show that $(-1, \sigma_i)_{v_i} = -1$ for every i. By Proposition 3.30,
$$(-1, \sigma_i)_{v_i} = (-1)^{v_i(\sigma_i) \frac{p^{f(\mathfrak{p}_i, p)} - 1}{2}}.$$
Thus we have to show that $f(\mathfrak{p}_i, p)$ and $v_i(\sigma_i)$ are odd. (Recall that $p \equiv -1 \mod 4$ by our hypothesis.)

Let
$$f(\mathfrak{p}_i, \mathfrak{p}_j) = [\mathcal{O}_{K_i}/\mathfrak{p}_i : \mathcal{O}_{K_j}/\mathfrak{p}_j]$$
be the degree of the extension $\mathcal{O}_{K_i}/\mathfrak{p}_i$ over $\mathcal{O}_{K_j}/\mathfrak{p}_j$ for $i > j$. Then since K_i is a quadratic extension of K_{i-1} and $\mathfrak{p}_{i-1}$ splits into two distinct primes in K_i, $f(\mathfrak{p}_i, \mathfrak{p}_{i-1}) = 1$ by Lemma 3.32 (1). Thus, $f(\mathfrak{p}_i, p) = 1$ by Lemma 3.32 (2).

Since $\sigma_0 = p(4a + p)$ and $p \neq a$, $v_0(\sigma_0) = 1$. For $i \geq 1$, first note that
$$[(K_i)_{v_i} : (K_{i-1})_{v_{i-1}}] = e(\mathfrak{p}_i, \mathfrak{p}_{i-1}) = 1.$$
Therefore
$$(K_i)_{v_i} = (K_{i-1})_{v_{i-1}} = \cdots = (K_0)_{v_0} = \mathbb{Q}_p$$
and v_i on $(K_i)_{v_i}$ is the p-adic valuation on $\mathbb{Q}_p$. Viewing $\sqrt{m_i}$ as an element of $\mathbb{Q}_p$, we can write $\sqrt{m_i} \equiv kp + 2a \mod p^2$ since $\sqrt{m_i} - 2a \equiv 0 \mod \mathfrak{p}_{i+1}$. By an induction, we can show that $k \equiv 4^{-i} \mod p$. Indeed, by squaring the above equation, we obtain
$$a(2a + \sqrt{m_{i-1}}) = m_i \equiv 4akp + 4a^2 \mod p^2,$$
and by the induction hypothesis for $i-1$, the conclusion for i follows. Thus
$$\sigma_i = m_i - 4a^2 \equiv 4^{-i+1}ap \not\equiv 0 \mod p^2.$$
This shows $v_i(\sigma_i) = 1$. It completes the proof for the special case $r = 2^k$.

Now we consider the general case. Given $r \geq 1$, write $r = 2^i s$ where s is an odd integer. In addition to the notations used in the previous special case, let
$$K = \mathbb{Q}(z^{1/r} + z^{-1/r}),$$
$$L = \mathbb{Q}(z^{1/r}) = K(\sqrt{\sigma})$$

where
$$\sigma = (z^{1/r} - z^{-1/r})^2 = (z^{1/r} + z^{-1/r})^2 - 4 \in K.$$
Then we have the following field extensions:

$$
\begin{array}{ccc}
 & & L \\
\mathfrak{p} \cdots & K & | \\
| & | & L_i \\
\mathfrak{p}_i \cdots & K_i & \\
| & | & \\
p \cdots & K_0 = \mathbb{Q} &
\end{array}
$$

We need to find a prime $\mathfrak{p}$ of K such that $(-1,\sigma)_{v_\mathfrak{p}} = -1$ for the valuation $v_\mathfrak{p}$ associated to $\mathfrak{p}$; in other words, both $v_\mathfrak{p}(\sigma)$ and $f(\mathfrak{p},p)$ must be odd by Proposition 3.30. Indeed we will find $\mathfrak{p}$ which is over the prime $\mathfrak{p}_i$ of K_i constructed in the above special case. Basically the existence of such a prime $\mathfrak{p}$ is guaranteed by a parity argument based on the fact that K is an *odd* degree extension over K_i. For this purpose we use Lemma 3.35. In our case it follows that

$$(*) \quad \begin{aligned} \sum_{\mathfrak{p}|\mathfrak{p}_i} f(\mathfrak{p},\mathfrak{p}_i) v_\mathfrak{p}(\sigma) &= \sum_{\mathfrak{p}|\mathfrak{p}_i} v_i(N^{K_{v_\mathfrak{p}}}_{(K_i)_{v_i}}(\sigma)) \\ &= v_i\Big(\prod_{\mathfrak{p}|\mathfrak{p}_i} N^{K_{v_\mathfrak{p}}}_{(K_i)_{v_i}}(\sigma) \Big) = v_i(N^K_{K_i}(\sigma)). \end{aligned}$$

Now $N^K_{K_i}(\sigma)$ can be computed as follows. It is easily seen that $[L:K] \leq 2$ and $[K:K_i] \leq s$. Since $v_i(\sigma_i) = 1$, $\sqrt{\sigma_i}$ is not contained in K_i and $[L_i:K_i] = 2$. From the irreducibility of $\lambda(t^{2^i s})$ and $\lambda(t^{2^i})$, $[L:L_i] = s$ and so
$$2s = [L:K_i] = [L:K][K:K_i].$$
This shows $[K:K_i] = s$ and $[L:K] = 2$. From this we obtain
$$\begin{aligned} N^K_{K_i}(\sigma) &= N^K_{K_i}((z^{\frac{1}{r}} - z^{-\frac{1}{r}})^2) \\ &= N^K_{K_i}(-N^L_K(z^{\frac{1}{r}} - z^{-\frac{1}{r}})) \\ &= -N^L_{K_i}(z^{\frac{1}{r}} - z^{-\frac{1}{r}}) = -N^{L_i}_{K_i} N^L_{L_i}(z^{\frac{1}{r}} - z^{-\frac{1}{r}}). \end{aligned}$$

The conjugates of $z^{1/r}$ over L_i are $z^{1/r}\zeta^k$ for $k = 0, 1, \ldots, s-1$ where ζ is a primitive s-th root of unity. Thus

$$\begin{aligned} N^L_{L_i}(z^{\frac{1}{r}} - z^{-\frac{1}{r}}) &= \prod_{k=0}^{s-1}(z^{\frac{1}{r}}\zeta^k - z^{-\frac{1}{r}}\zeta^{-k}) \\ &= \zeta^{\frac{s(s-1)}{2}} \prod_{k=0}^{s-1}(z^{\frac{1}{r}} - z^{-\frac{1}{r}}\zeta^{-2k}) \\ &= z^{\frac{1}{2^i}} - z^{-\frac{1}{2^i}} = \sqrt{\sigma_i} \end{aligned}$$

since s is odd. Therefore
$$N^K_{K_i}(\sigma) = -N^{L_i}_{K_i}(\sqrt{\sigma_i}) = \sigma_i.$$

Now $(*)$ becomes
$$\sum_{\mathfrak{p}|\mathfrak{p}_i} f(\mathfrak{p},\mathfrak{p}_i) v_\mathfrak{p}(\sigma) = v_i(\sigma_i) = 1.$$

By a parity argument, it follows that there exists at least one prime $\mathfrak{p}$ over $\mathfrak{p}_i$ such that both $f(\mathfrak{p},\mathfrak{p}_i)$ and $v_\mathfrak{p}(\sigma)$ are odd. It completes the proof. □

3.5.3. Structure of $\mathcal{G}_n$. Fix $c > 0$. Since there are infinitely many primes $a \equiv 1 \mod 4$, we can construct infinitely many Seifert matrices A_i such that $\mathcal{A}_i = \phi_c(A_i)$ has order 2 in $\mathcal{G}_n$ using Theorem 3.36. Similarly, since there are infinitely many pairs (a,p) satisfying the condition of Theorem 3.37 (e.g., first choose a prime $p \equiv -1 \mod 4$ and choose sufficiently large prime a), we can construct infinitely many Seifert matrices B_i such that $\mathcal{B}_i = \phi_c(B_i)$ has order 4 in $\mathcal{G}_n$. Furthermore, we can show the following result:

THEOREM 3.38. *The subgroup H generated by the $\mathcal{A}_i$ and $\mathcal{B}_i$ is isomorphic to $(\mathbb{Z}/2)^\infty \oplus (\mathbb{Z}/4)^\infty$ and is a summand of (the torsion subgroup of) $\mathcal{G}_n$.*

PROOF. Since (the irreducible decompositions of) the Alexander polynomials of the A_i and B_i are distinct, the nontrivial primary parts of the A_i and B_i are "orthogonal" in the following sense: let $z_i \ne -1$, $w_i \ne -1$ be zeros of $\Delta_{A_i}(t)$ and $\Delta_{B_i}(t)$, respectively. Then the z_i and w_i are mutually distinct complex numbers such that the z-primary part of A_j (resp. B_j) is trivial for all $z \in \{z_i, w_i\}$ but $z = z_j$ (resp. w_j)

By Corollary 3.19, there exist $\alpha_i, \beta_i \in P$ such that $(\alpha_i)_c = z_i$, $(\beta_i)_c = w_i$. Combining the above orthogonality with the computation in the proofs of Theorems 3.36 and 3.37, we have the following properties:

(1) For $\alpha \in \{\alpha_i, \beta_i\}$, $e_\alpha(\mathcal{A}_i)$ is nontrivial if and only if $\alpha = \alpha_i$, and $e_\alpha(\mathcal{B}_i)$ is nontrivial if and only if $\alpha = \beta_i$.
(2) $d_{\beta_i}(2\mathcal{B}_j)$ is nontrivial if and only if $i = j$.

Suppose that $\sum a_i \mathcal{A}_i + \sum b_i \mathcal{B}_i = 0$, where all but finitely many a_i and b_i are zero. Taking e_{α_i} of both sides and using the property (1) above, we obtain
$$0 \equiv a_i \cdot e_{\alpha_i}(\mathcal{A}_i) = a_i \mod 2$$
for each i. Similarly, taking e_{β_i}, it is shown that b_i is even for each i. Since $\mathcal{A}_i$ has order two, $a_i \mathcal{A}_i = 0$ so that the relation becomes $\sum b'_i(2\mathcal{B}_i) = 0$ where $b_i = 2b'_i$. Taking d_{β_i} of both sides and using the property (2) above, it follows that b'_i is even, i.e., b_i is a multiple of 4 for each i. Since $\mathcal{B}_i$ has order four, $b_i \mathcal{B}_i = 0$. This proves that $H \cong (\mathbb{Z}/2)^\infty \oplus (\mathbb{Z}/4)^\infty$.

To show that H is a summand of the torsion subgroup T of $\mathcal{G}_n$, we appeal to the following result from group theory: a subgroup H of an abelian group G is called a *pure subgroup* if $kG \cap H \subset kH$ for all k.

3.5. COMPUTATION OF $d(\mathcal{A})$

LEMMA 3.39. *Suppose G is an abelian group and H is a pure subgroup of G. If $rH = 0$ for some $r > 0$, then H is a direct summand of G.*

For a proof, see [**41**, p. 199].

In our case, $4T = 0$ and hence $4H = 0$. To verify that our H is a pure subgroup of T, i.e., $kT \cap H \subset kH$ for all k, we write $k = 2^s \cdot k'$, where k' is odd. Since $4H = 0 = 4T$, $k'T = T$ and $k'H = H$. Thus $kT = 2^s T$ and $kH = 2^s H$. Again appealing to $4T = 0 = 4H$, we may assume that $s = 1$, i.e., it suffices to check $2T \cap H \subset 2H$. Suppose $\sum a_i \mathcal{A}_i + \sum b_i \mathcal{B}_i = 2\mathcal{A}$ where $\mathcal{A} \in T$. As before, by taking e_{α_i} and e_{β_i}, a_i and b_i are even. Hence

$$\sum a_i \mathcal{A}_i + \sum b_i \mathcal{B}_i = 2\big(\sum b'_i \mathcal{B}_i\big) \in 2H$$

where $b_i = 2b'_i$. This completes the proof. $\square$

COROLLARY 3.40. *$\mathcal{G}_n$ is isomorphic to $\mathbb{Z}^\infty \oplus (\mathbb{Z}/2)^\infty \oplus (\mathbb{Z}/4)^\infty$.*

PROOF. Obviously $\mathcal{G}_n$ is the direct sum of its torsion subgroup T and the free abelian group $\mathcal{G}_n/T$. It is already known that the rank of $\mathcal{G}_n/T$ is infinite (e.g., see [**7**]). Since $4T = 0$, T is a direct sum of cyclic groups of order 2 and 4, by Prüfer's theorem (e.g., see [**41**, p. 197]). Combining this with Theorem 3.38, the conclusion follows. $\square$

EXAMPLE 3.41. We give concrete examples of Seifert matrices representing finite order elements in $\mathcal{G}_n$. The Alexander polynomial described in Theorem 3.36 can be realized by a Seifert matrix by Theorem 3.22. In fact, by the algorithm used in the proof of Theorem 3.22, we obtain the following Seifert matrix:

$$\begin{bmatrix} -a & 1 \\ 0 & 1 \end{bmatrix}, \quad \text{if } \epsilon = 1,$$

$$\begin{bmatrix} 0 & 1 & a & 0 \\ 0 & 0 & 1 & 0 \\ -a & -1 & -1 & 0 \\ 0 & 0 & 0 & 1 \end{bmatrix}, \quad \text{if } \epsilon = -1.$$

Therefore the image of this matrix under $G_n = G_{n,c} \to \mathcal{G}_n$ is of order two.

In a similar way, we obtain a Seifert matrix

$$\begin{bmatrix} -\frac{a}{p} & 1 \\ 0 & 1 \end{bmatrix}, \quad \text{if } \epsilon = 1,$$

$$\begin{bmatrix} 0 & 1 & \frac{a}{p} & 0 \\ 0 & 0 & 1 & 0 \\ -\frac{a}{p} & -1 & -1 & 0 \\ 0 & 0 & 0 & 1 \end{bmatrix}, \quad \text{if } \epsilon = -1,$$

whose Alexander polynomial is as described in Theorem 3.37, so that its image under $G_n = G_{n,c} \to \mathcal{G}_n$ is of order four.

We obtain infinitely many matrices by choosing different values of a and p in the above construction, and by Theorem 3.38, the elements in

$\mathcal{G}_n$ represented by these matrices generate a summand of $\mathcal{G}_n$ isomorphic to $(\mathbb{Z}/2)^\infty \oplus (\mathbb{Z}/4)^\infty$.

REMARK 3.42. As a consequence of the results of this section, we can compute some surgery obstruction Γ-groups. First we rephrase our computation in terms of relative Witt groups. From Theorem 3.22, the group $G_{n,c}$ injects into the Witt group $W_\epsilon(\mathbb{Q}[\mathbb{Z}], S')$ where $S' = \{f(t) \in \mathbb{Q}[\mathbb{Z}] \mid f(1) \neq 0\}$. Consider a direct system consisting of $W_i = W_\epsilon(\mathbb{Q}[\mathbb{Z}], S')$ and homomorphisms $W_i \to W_{ri}$ induced by $t \to t^r$. Then our arguments in this section show that
$$\varinjlim W_i \cong \mathbb{Z}^\infty \oplus (\mathbb{Z}/2)^\infty \oplus (\mathbb{Z}/4)^\infty.$$

On the other hand, every finitely generated $\mathbb{Q}[\mathbb{Z}]$-module has homological dimension one, since $\mathbb{Q}[\mathbb{Z}]$ is a PID. Thus the relative Witt group $W_\epsilon(\mathbb{Q}[\mathbb{Z}], S')$ can be identified with the relative L-group $L_{n+3}(\mathbb{Q}[\mathbb{Z}], S')$, which is easily seen to be isomorphic to $L_{n+3}(\mathbb{Q}[\mathbb{Z}], S_0)$ where $S_0 = 1 + \operatorname{Ker}\varepsilon$ and $\varepsilon \colon \mathbb{Q}[\mathbb{Z}] \to \mathbb{Q}$ is the augmentation map. From exact sequences relating Γ-groups, relative L-groups, and localizations (e.g., see Ranicki's book [**40**]), we have

$$\Gamma_{n+3}\begin{pmatrix}\mathbb{Q}[\mathbb{Z}] \xrightarrow{\text{id}} \mathbb{Q}[\mathbb{Z}] \\ \text{id}\downarrow \qquad \downarrow \varepsilon \\ \mathbb{Q}[\mathbb{Z}] \xrightarrow{\varepsilon} \mathbb{Q}\end{pmatrix} \cong L_{n+3}(\mathbb{Q}[\mathbb{Z}], S_0) \cong W_i.$$

Taking limits, it follows that

$$\Gamma_{n+3}\begin{pmatrix}\mathbb{Q}[\mathbb{Q}] \xrightarrow{\text{id}} \mathbb{Q}[\mathbb{Q}] \\ \text{id}\downarrow \qquad \downarrow \varepsilon \\ \mathbb{Q}[\mathbb{Q}] \xrightarrow{\varepsilon} \mathbb{Q}\end{pmatrix} \cong \mathbb{Z}^\infty \oplus (\mathbb{Z}/2)^\infty \oplus (\mathbb{Z}/4)^\infty.$$

This Γ-group is closely related to Cochran and Orr's homology surgery theoretic approach to rational knot concordance.

CHAPTER 4

Geometric structure of $\mathcal{C}_n$

4.1. Realization of rational Seifert matrices

The aim of this section is to prove the following realization theorem of rational Seifert matrices. As before we adopt the convention $\epsilon = (-1)^{q+1}$.

THEOREM 4.1. *Suppose A is a rational square matrix and c is a positive integer. Then A is a Seifert matrix of complexity c for some $(2q-1)$-knot in a rational sphere bounding a parallelizable rational ball if and only if there is a rational square matrix P such that $P(A - \epsilon A^T)P^T$ is an integral unimodular (i.e., invertible over $\mathbb{Z}$) matrix with even diagonal entries. In addition, we require $\operatorname{sign}(A + A^T) \equiv 0 \mod 16$ when $q = 2$.*

We remark that in contrast to the integral case, the "even" condition gives further restriction when $\epsilon = -1$ since P and A have rational entries. It can be omitted when $\epsilon = 1$. In the topological category (where submanifolds are assumed to be locally flat), the signature condition for $q = 2$ is not required.

The only if part was already discussed. For the if part, we may assume that $Q = A - \epsilon A^T$ is an integral unimodular matrix with even diagonals by replacing A by PAP^T. We will describe a concrete construction of a rational knot equipped with a generalized Seifert surface of complexity c whose Seifert matrix is A.

4.1.1. Special case: complexity one. First we consider the special case $c = 1$. Since Q has even diagonal entries, we can choose an integral matrix B such that $Q = B - \epsilon B^T$. Indeed, denoting $Q = (q_{ij})$ and $B = (b_{ij})$, we can choose b_{ij} arbitrarily for $i < j$, and then, let $b_{ii} = q_{ii}/2$ and $b_{ji} = q_{ji} + \epsilon b_{ij}$ for $i < j$.

By Levine [33], there is a Seifert surface E of a $(2q-1)$-dimensional knot in S^{2q+1} such that E consists of one 0-handle and $2g$ q-handles, and B is its (integral) Seifert matrix with respect to the basis $\{x_i\}$ of $H_q(E)$ where x_i is an embedded q-sphere in E obtained by attaching a q-disk in the 0-handle to the core of the i-th q-handle.

We will do surgery on S^{2q+1} along q-spheres in $S^{2q+1} - E$ so that E becomes a desired Seifert surface in a rational sphere. Write $a_{ij} - b_{ij} = m_{ij}/n_{ij}$, where $A = (a_{ij})$ and m_{ij} and n_{ij} are integers. Choose a collection of disjoint embedded q-spheres
$$\{c_{ij}^+, c_{ij}^- \mid 1 \leq i \leq j \leq 2g\}$$

in $S^{2q+1} - E$ such that

$$\text{lk}(c_{ij}^+, x_k) = \delta_{ik},$$
$$\text{lk}(c_{ij}^-, x_k) = \delta_{jk} m_{ij},$$
$$\text{lk}(c_{ij}^+, c_{kl}^+) = \text{lk}(c_{ij}^-, c_{kl}^-) = 0,$$
$$\text{lk}(c_{ij}^+, c_{kl}^-) = \begin{cases} \delta_{ik}\delta_{jl} \cdot n_{ij} & \text{for } i < j, \\ \delta_{ik}\delta_{jl} \cdot 2n_{ij} & \text{for } i = j, \end{cases}$$

where lk denotes the linking number in S^{2q+1} and δ_{ij} is the Kronecker delta. See the schematic picture in Figure 1. Let Σ be the result of surgery on S^{2q+1} along $\{c_{ij}^{\pm}\}$, where $c_{ij}^{\pm}$ is framed as follows. (We call it the *null-framing*.) Viewing S^{2q+1} as the boundary of D^{2q+2}, we can choose disjoint embedded $(q+1)$-disks $D_{ij}^{\pm}$ in D^{2q+2} which meet S^{2q+1} orthogonally at $c_{ij}^{\pm}$. Then the normal bundle of $D_{ij}^{\pm}$ in D^{2q+2} admits a unique trivialization (up to fiber homotopy) which induces a trivialization of the normal bundle of $c_{ij}^{\pm}$ in S^{2q+1}. Our Σ is the result of surgery along this framing on the $c_{ij}^{\pm}$.

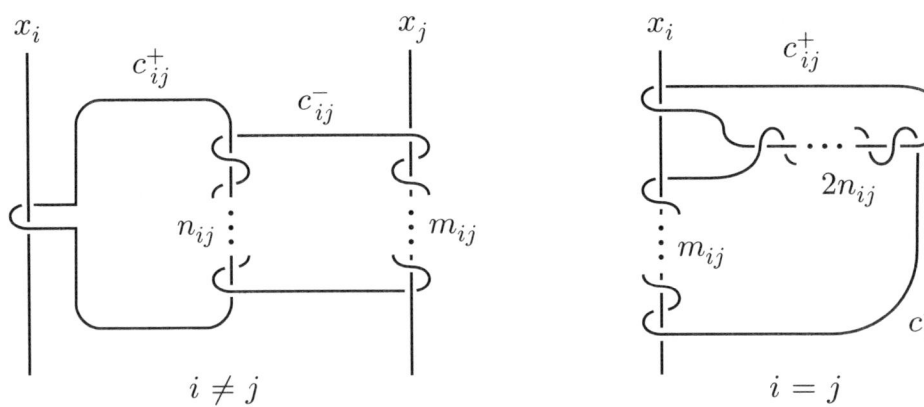

FIGURE 1

Now E becomes a Seifert surface F in Σ. We claim that

$$\text{lk}_{\Sigma}(x_i^+, x_j) = \text{lk}(x_i^+, x_j) + \frac{m_{ij}}{n_{ij}}$$

where x_i^+ is the q-sphere obtained by pushing x_i slightly along the positive normal direction of F and lk_{Σ} denotes the rational linking number in Σ. Then it follows that F has Seifert matrix A since

$$\text{lk}_{\Sigma}(x_i^+, x_j) = b_{ij} + \frac{m_{ij}}{n_{ij}} = a_{ij}.$$

To prove the claim, we appeal to the following lemma which is a higher dimensional version of [**7**, Theorem 3.1]:

4.1. REALIZATION OF RATIONAL SEIFERT MATRICES

LEMMA 4.2. *Suppose $K_1, \cdots, K_m$ are disjoint framed q-spheres embedded in S^{2q+1} such that surgery on S^{2q+1} along the K_i produces a rational sphere Σ. For two disjoint q-cycles a and b in S^{2q-1} which are disjoint to the K_i, the linking number of a and b in Σ is*

$$\operatorname{lk}_\Sigma(a,b) = \operatorname{lk}_{S^{2q+1}}(a,b) - x^T L^{-1} y$$

where $x = (a_i)$ and $y = (b_i)$ are column vectors given by $a_i = \operatorname{lk}_{S^{2q+1}}(a, K_i)$ and $b_i = \operatorname{lk}_{S^{2q+1}}(b, K_i)$ and L is the linking matrix whose (i,j)-entry is the linking number of K_i and the preferred parallel of K_j obtained by pushing K_j slightly along the given framing.

The special case of $q = 1$ was proved in [**7**, Theorem 3.1]. Since the same argument also works for any q, we omit the details.

Returning to the proof of Theorem 4.1, we apply Lemma 4.2 to compute the linking of x_i^+ and x_j in Σ. Our linking matrix L is the block sum of the following 2×2 matrices representing the linking of $\{c_{kl}^+, c_{kl}^-\}$:

$$\begin{bmatrix} 0 & n_{kl} \\ n_{kl} & 0 \end{bmatrix} \text{ for } k < l, \quad \begin{bmatrix} 0 & 2n_{kl} \\ 2n_{kl} & 0 \end{bmatrix} \text{ for } k = l.$$

L^{-1} is the block sum of their inverses:

$$\begin{bmatrix} 0 & 1/n_{kl} \\ 1/n_{kl} & 0 \end{bmatrix} \text{ for } k < l, \quad \begin{bmatrix} 0 & 1/2n_{kl} \\ 1/2n_{kl} & 0 \end{bmatrix} \text{ for } k = l.$$

By our choice of $c_{kl}^\pm$, the only nontrivial contribution of the $x^T L^{-1} y$ term of the formula in Lemma 4.2 is from the block associated to $\{c_{ij}^+, c_{ij}^-\}$. Indeed $\operatorname{lk}_\Sigma(x_i^+, x_j)$ is given by

$$b_{ij} - \begin{bmatrix} 1 & 0 \end{bmatrix} \begin{bmatrix} 0 & -1/n_{ij} \\ -1/n_{ij} & 0 \end{bmatrix} \begin{bmatrix} 0 \\ m_{ij} \end{bmatrix} \quad \text{for } i = j,$$

$$b_{ij} - \begin{bmatrix} 1 & m_{ij} \end{bmatrix} \begin{bmatrix} 0 & -1/2n_{ij} \\ -1/2n_{ij} & 0 \end{bmatrix} \begin{bmatrix} 1 \\ m_{ij} \end{bmatrix} \quad \text{for } i = j.$$

In both cases it is equal to $b_{ij} + m_{ij}/n_{ij}$, as desired. This proves the claim and completes the construction for $c = 1$.

EXAMPLE 4.3. We again consider the matrix

$$A = \begin{bmatrix} -\frac{a}{p} & 1 \\ 0 & 1 \end{bmatrix}$$

which represents a 4-torsion element in $\mathcal{G}_n$ for $n \equiv 1 \mod 4$. The algorithm described above gives us a rational knot K which has Seifert matrix A. Figure 2 illustrates K for $n = 1$. We have an obvious Seifert surface F of K with one 0-handle and two middle-dimensional handles. One middle dimensional handle of F is twisted once so that the core has self-linking number 1. Another handle of F is untwisted so that the core has vanishing self-linking number. There are a pairs of null-framed surgery spheres where each pair has linking number $2p$ and each sphere has linking number 1 with the latter handle of F. The ambient space of K is obtained by performing surgery along these $2a$ spheres. Figure 2 can also be viewed as a schematic picture of a rational knot in higher odd dimensions, which has order 4 in $\mathcal{C}_n$.

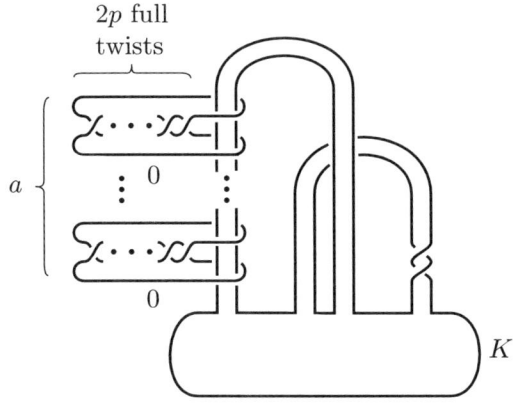

FIGURE 2

4.1.2. General case. Suppose $c > 1$. Given A, from the above special case, there is a knot K' in a rational $(2q+1)$-sphere Σ' equipped with a generalized Seifert surface F' of complexity 1 whose Seifert matrix is A. We apply the construction of [**7**, p. 1179–1180] to produce a generalized Seifert surface of complexity c: choose an embedded $(2q-1)$-sphere K bounding a $2q$-ball B disjoint to K' in Σ', and choose a simple closed curve C in $\Sigma' - (K' \cup K)$ which meets B and F' transversally at a negative intersection point and c positive intersection points so that the linking numbers of C with K and K' are -1 and c, respectively. See the schematic picture in Figure 3 (a).

By our construction of Σ' in the special case above, we can view Σ as a result of surgery on S^{2q+1} so that the "null-framing" on C and K' are defined. Note that F' induces the null-framing of K'. We perform null-framed surgery on Σ along C and a parallel K' which is taken with respect to the null-framing. The result is again a rational $(2q+1)$-sphere, which we denote by Σ. We can view K as a knot in Σ.

A generalized Seifert surface of K is constructed as follows. Consider the union of F' and c parallel copies of B. Puncturing it at the intersection with C and attaching c pipes, we obtain a submanifold F'' in Σ' bounded by K' and c parallel copies of K. Since F' induces the null-framing on K', there exists a $2q$-disk in $\Sigma - \text{int}(F'')$ bounded by K'. Attaching this disk to F'', we obtain a generalized Seifert surface F of complexity c for K. See Figure 3 (b).

F and F' have the same H_q (or $\text{Coker}\{H_q(\partial -) \to H_q(-)\}$ if $q = 1$). For any q-cycles x and y on F', the linking numbers $\text{lk}_\Sigma(x^+, y)$ in Σ and

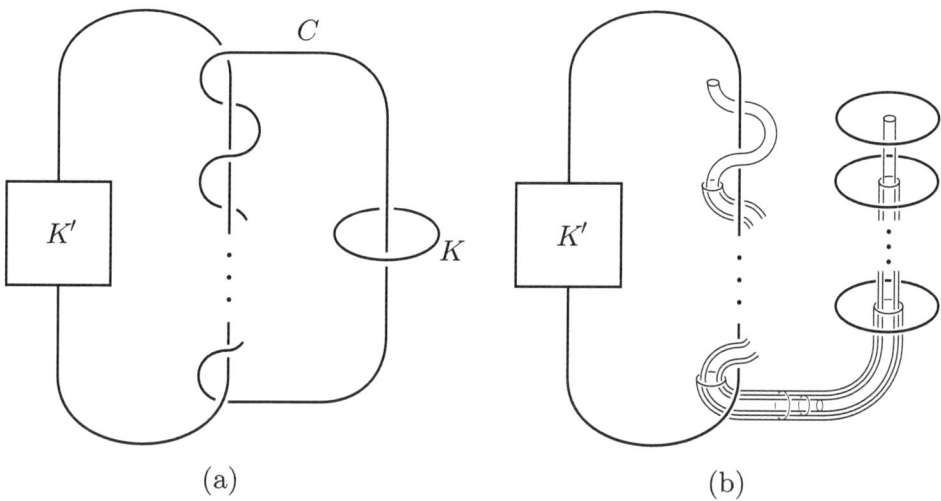

FIGURE 3

$\text{lk}_{\Sigma'}(x^+, y)$ in Σ' are the same. Indeed for $q > 1$, since $H_q(\Sigma' - (C \cup K'); \mathbb{Q}) = 0$, we can choose a $(q+1)$-chain u in $\Sigma' - (C \cup K)$ such that $\partial u = rx^+$ for some $r > 0$, and then both $\text{lk}_\Sigma(x^+, y)$ and $\text{lk}_{\Sigma'}(x^+, y)$ are equal to $(1/r)(u \cdot y)$. For $q = 1$, using the fact that both x^+ and y have linking number zero with C and K', we can apply Lemma 4.2. This shows that A is a Seifert matrix of F.

The only remaining thing to verify is that our ambient space Σ bounds a rational ball with stably trivial normal bundle. For this purpose we think of the trace of the surgery giving Σ, and perform surgery on the interior of the trace to obtain a desired rational ball. Details are as follows. From the fact that Σ is obtained by performing null-framed surgery on S^{2q+1}, we can see that a framed $(2q+2)$-manifold W with boundary Σ is obtained by attaching to B^{2q+2} a $2q$-handle, a 2-handle, and even number of $(q+1)$-handles. Note that the $(q+1)$-handles give rise to a symplectic basis of $H_{q+1}(W; \mathbb{Q})$, with respect to the intersection form, by the above construction of Σ. Standard surgery techniques shows that we can perform framed surgery on W to kill the homology classes of W represented by the 2-handle and half of the $(q+1)$-handles forming a Lagrangian (they are all spherical obviously). An alternative ad-hoc method to see this is as follows: the union of B^{2q+1} and the concerned handles (the other handles are ignored) is embedded in $S^{2q+1} = \partial B^{2q+2}$ and the spheres along which we want to do surgery bound disjoint disks in B^{2q+2}. So we can do null-framed surgery. The result is a rational ball Δ which is framed and has boundary Σ.

4.2. Construction of slice disks in rational balls

In this section we study the subgroup $b\mathcal{C}_n$ in $\mathcal{C}_n$ generated by concordance classes of knots in rational spheres bounding a parallelizable rational ball.

We start with some preliminary lemmas. First, the following result will be used to simplify rational balls.

LEMMA 4.4. *Suppose $q > 2$ and Σ is a rational $(2q-1)$-sphere bounding a parallelizable rational ball. Then Σ bounds a parallelizable rational ball which is $(q-2)$-connected.*

This can be proved by applying standard techniques of surgery. It suffices to perform surgery below the middle dimension, and hence we have no nontrivial obstruction. What follows is (a sketch of) a proof using framed surgery.

PROOF. Suppose $\partial W = \Sigma$, W is a parallelizable rational $2q$-ball. Then we can perform framed surgery on the interior of W, below dimension $q-1$, to construct a parallelizable $(q-2)$-connected $2q$-manifold V bounded by Σ. By surgery killing the homology class α of an embedded i-sphere ($i \leq q-2$), $H_{i+2}, \ldots, H_q$ are left unchanged; H_{i+1} is left unchanged if and only if α is of infinite order in H_i; the rank of H_{i+1} increases if α is torsion. Hence $H_q(V) = H_q(W)$ but $H_{q-1}(V)$ may have nontrivial free part. By surgery again, we can kill the generators of the free part of $H_{q-1}(V)$ keeping $H_q(V)$ unchanged. This gives us a desired rational ball. □

A similar argument proves the following result for even dimensional rational spheres:

LEMMA 4.5. *Suppose $q > 1$ and Σ is a q-parallelizable rational $2q$-sphere (i.e., the restriction of the normal bundle on the q-skeleton is stably trivial). In addition, if q is odd, suppose Σ has vanishing Arf invariant. Then Σ bounds a rational ball which is $(q-1)$-connected.*

PROOF. First we perform framed surgery on Σ to make it a PL $2q$-sphere. Capping the trace of surgery with a $(2q+1)$-ball, we obtain a q-parallelizable $(2q+1)$-manifold W bounded by Σ. By framed surgery on the interior of W killing some i-spheres, $i \leq q-1$, we may assume that W is $(q-1)$-connected.

Now we do surgery on W to kill the free part of $H_q(W)$. Let α be an embedded q-sphere in W representing an infinite order class in $H_q(W)$. Since ∂W is a rational sphere, from the duality with rational coefficients it follows that the natural homomorphism $H_{q+1}(W) \to H_{q+1}(W, W - \alpha) \cong \mathbb{Z}$, which is given by the intersection with α, is a nontrivial map. Its image is an ideal generated by a positive integer c. It can be seen that surgery along α kills the homology class of α but introduces a new order c element to $H_q(W)$ (e.g. see [**27**, Lemma 5.6]). Repeating this, we can kill the free part of $H_q(W)$ (but the torsion part may grow). This gives a rational ball bounded by Σ. □

Consider the following codimension one ambient surgery problem: suppose that M is an m-manifold embedded in the boundary of an $(m+2)$-manifold W, α is an embedded i-sphere in M, and δ is an properly embedded

$(i+1)$-disk in W bounded by α. When can one do ambient surgery on M using the disk δ? In other words, when can one obtain an $(i+1)$-handle attached on M by thickening δ? The following result is proved by well-known arguments (c.f., [**2**, p. 86], [**33**, p. 235]).

LEMMA 4.6. *There is an obstruction $o \in \pi_i(S^{m-i})$ which vanishes if and only if we can do ambient surgery along α on M using δ in W.*

PROOF. The normal bundle ξ of $\delta \subset W$ can be identified with $\delta \times D^{m-i+1}$ in a unique way, being a bundle over a contractible space. The associated sphere bundle restricted on α is a trivial sphere bundle $\alpha \times S^{m-i}$. By restricting on α the positive normal direction of M in ∂W, which is uniquely determined by the orientations, we obtain a section $\alpha \to \alpha \times S^{m-i}$, which gives rise to an element $o \in \pi_i(S^{m-i})$.

If o is trivial, the section $\alpha \to \alpha \times S^{m-i}$ extends to $\delta \to \delta \times S^{m-i}$, and the orthogonal complement of this direction in $\xi \cong \delta \times D^{m-i+1}$ gives us an $(i+1)$-handle that can be used to do surgery on M along α. The converse is proved in a similar way. □

The followings are consequences of (the proof of) the above lemma.

LEMMA 4.7.
(1) *If $2i < m$, we can always do ambient surgery along α on M using δ in W.*
(2) *If $2i = m$ and W is a rational ball, the obstruction $o \in \pi_i(S^i) = \mathbb{Z}$ is given by the linking number of α and a pushoff of α along the positive normal direction of M in the rational sphere ∂W.*

4.2.1. Slicing odd dimensional rational knots. In this subsection we discuss how to construct a slice disk in a rational ball for an odd-dimensional rational knot. First we focus on the special case of knots of complexity 1, i.e., knots bounding a Seifert surface. We will call such knots *primitive*, following [**7**]. The below proposition reduces the problem into the case of simple knots: we call a primitive $(2q-1)$-knot *simple* if it bounds a $(q-1)$-connected Seifert surface.

PROPOSITION 4.8. *Suppose K is a primitive $(2q-1)$-knot in a rational $(2q+1)$-sphere Σ bounding a parallelizable rational ball. Then K is concordant to a primitive simple knot in a rational sphere bounding a parallelizable rational ball. In addition, they are concordant via a concordance of complexity 1.*

Although its statement is very similar to a corresponding result of [**33**] for integral knots, the proof of Proposition 4.8 requires more sophisticated arguments since a rational $(2q+2)$-ball bounded by Σ may have nontrivial homotopy groups even below the middle dimension. See the latter half of the proof below.

PROOF. For $q = 1$, the conclusion is obvious. Suppose $q > 1$. Let denote by Δ a parallelizable rational ball with boundary Σ. The first part of the proof is similar to an argument of [**33**] for ordinary knots; we do ambient surgery on a Seifert surface F of K, in the rational ball Δ, to obtain a $(q-1)$-connected submanifold in Δ. If we were doing abstract surgery, it would suffice to do surgery on F along suitable disjoint spheres of dimension $\leq (q-1)$. To do this in the ambient space Δ, first we assume that Δ is $(q-1)$-connected by appealing to Lemma 4.4. Then we can choose immersed disks of dimension $\leq q$ in Δ which are bounded by these spheres, and by general position, we can assume that these disks are embedded and mutually disjoint since Δ is $(2q+2)$-dimensional. Then by appealing to Lemma 4.7 (1), we can do ambient surgery on F using these disks. The trace of surgery is a 2-sided $(2q+1)$-submanifold W in Δ which is a cobordism, relative to the boundary, between F and a $(q-1)$-connected $2q$-manifold F'.

We remark that, in the case of ordinary knots, $\Delta = B^{2q+2}$ and it is able to find a honest ball in the interior of Δ whose intersection with W is F', using an engulfing technique as in [**33**], so that $\partial F'$ is a desired simple knot in the boundary of the honest ball. In contrast to this, in our case, such a ball may not exist. The best we can do is construct a rational ball instead. The remaining part of our proof is devoted to this construction.

Let V be Δ cut along W. Then, by a general position argument, $\pi_i(V) \cong \pi_i(\Delta)$ for $i \leq q$, since W is obtained by attaching handles of index $\leq q$ to $F \times [0,1]$. In particular, $\pi_q(V) \cong \pi_q(\Delta) \cong H_q(\Delta)$ is finite. Let $r = |\pi_q(V)|$. Note that F' consists of a 0-handle and $2g$ q-handles by handle theory. For each q-handle, choose an immersed q-sphere in F' representing r times the generator of $H_q(F')$ represented by the q-handle. While we can assume that each of these q-spheres is embedded by isotopy, two different q-spheres may meet at several points. Let X be the union of these q-spheres. Then we may assume that X has the homotopy type of

$$(\bigvee^{2g} S^q) \vee (\bigvee^m S^1).$$

We will construct a complex Y with the homotopy type of $(\bigvee^{2g} S^q)$ by attaching 2-disks to X killing the S^1-factors. Let N be a regular neighborhood of X in F'. Since $q \geq 2$,

$$\pi_1(\partial N) \cong \pi_1(N - X) \longrightarrow \pi_1(N) \cong \pi_1(X)$$

is surjective. Thus we can choose disjoint circles γ_k on ∂N representing generators of $\pi_1(X)$. We claim that there are disjoint embedded 2-disks in $F' - \text{int } N$ bounded by the γ_k. Then the union of N and these 2-disks is a desired complex Y. If $q \geq 3$, the claim follows from $\pi_1(F' - X) \cong \pi_1(F') = 0$.

For $q = 2$, we need more sophisticated ad-hoc arguments. As done in [**33**, p. 235], by taking connected sum with some copies of $S^2 \times S^2$ in the ambient space Δ, we may assume that F' is homeomorphic to $\#^g S^2 \times S^2$ with a puncture. We can view F' as a handlebody with a 0-handle B^4 and $2g$ 2-handles attached along a split union of Hopf links $K_i^1 \cup K_i^2$ ($i = 1, \ldots, g$) contained in ∂B^4. We can choose r-punctured spheres C_i^1 and C_i^2 properly

4.2. CONSTRUCTION OF SLICE DISKS IN RATIONAL BALLS

embedded in B^4 such that C_i^j is bounded by the union of r parallel copies of K_i^j, $C_i^j \cap C_{i'}^{j'} = \emptyset$ for $i \neq i'$, and $C_i^1 \cap C_i^2$ consists of r^2 points. Figure 4 illustrates the configuration of C_i^1 and C_i^2 as a "movie" along the radial direction of B^4, i.e., the intersection of C_i^j with the level sphere $\{x \in \mathbb{R}^4 : |x| = t\}$, viewing B^4 as the unit ball in $\mathbb{R}^4$. Attaching parallel copies of the cores of the 2-handles of F' to the union of all the C_i^j, we obtain a complex X which is homotopy equivalent to

$$(\bigvee^{2g} S^2) \vee (\bigvee^{r^2-1} S^1).$$

In Figure 4, the dotted lines represent $(r^2 - 1)$ 2-disks in $B^4 \subset F'$ whose boundaries are the concerned curves γ_k representing the S^1-factors of X. Attaching these 2-disks to X, we obtain the complex Y.

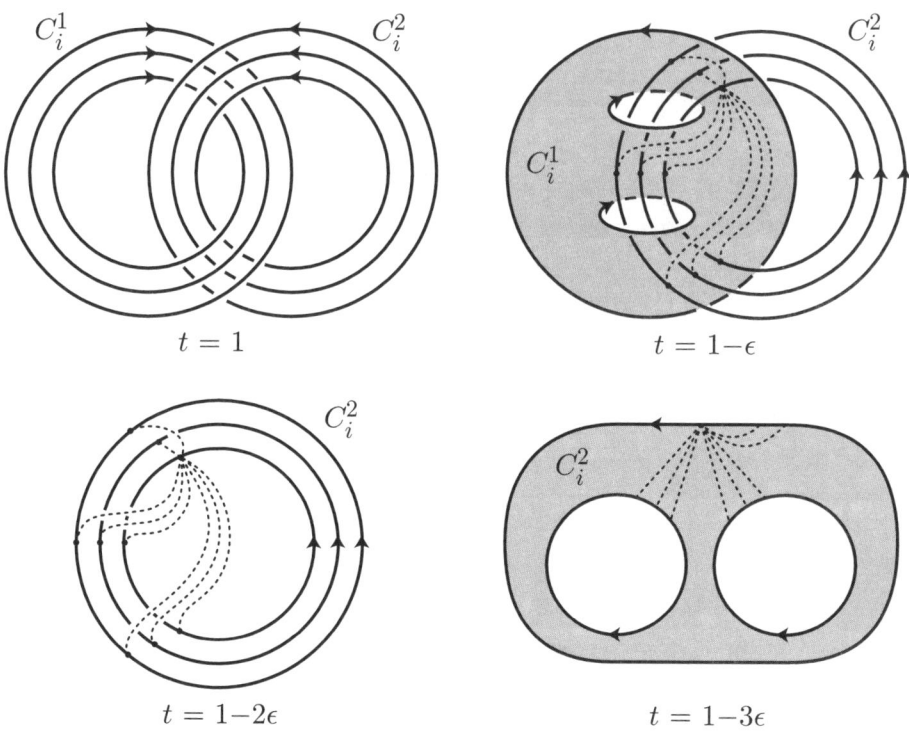

FIGURE 4

Now we use the complex Y to construct a rational ball whose boundary contains F'. Recall that W is the trace of ambient surgery producing F' from F. For notational convenience, we identify a bicollar of F' in W with $F' \times [0, 1]$, where $F' \subset \partial W$ is identified with $F' \times 0$. $Y \times 0 \subset F' \times 0 \subset V$ ($= \Delta$ cut along W) is null-homotopic in V by our choice of r. By the engulfing

theorem of Hirsch [**21**], there is a $(2q+2)$-cell C in V such that $Y \times 0 \subset \partial C$. Let
$$Z = C \cup (Y \times [0, \tfrac{1}{2}]) \cup (F' \times \tfrac{1}{2})$$
and view it as a subset of Δ. $H_*(Z)$ is trivial except $H_0(Z) = \mathbb{Z}$, $H_q(Z) = (\mathbb{Z}/r)^n$. Choose a regular neighborhood Δ' of Z in Δ such that $F' \times 1 \subset \partial \Delta'$. Δ' has the homotopy type of Z and hence is a rational ball. Being a codimension zero submanifold of Δ, Δ' is parallelizable. Now it is easily seen that K is concordant to the rational knot $\partial F' \times 1 \subset \partial \Delta'$, via the concordance
$$S^{2q-1} \times [0, 1] \cong \partial W - \text{int}(F) - F' \times [0, 1) \subset \Delta - \text{int}(\Delta'),$$
and $F' \times 1$ is a $(q-1)$-connected Seifert surface of $\partial F' \times 1$.

Finally, by the Thom–Pontryagin construction with the codimension one submanifold
$$W - F \times [0, 1) \subset \Delta - \text{int}(\Delta'),$$
we obtain an S^1-structure $E \to S^1$ of complexity 1 where E is the exterior of the concordance, that is, it induces a homomorphism $H_1(E)/\text{torsion} = \mathbb{Z} \to \mathbb{Z}$ sending a meridian to $1 \in \mathbb{Z}$. This shows that the concordance has complexity 1. $\square$

REMARK 4.9. Although we do not need it in this paper, the following generalization of Proposition 4.8 can be proved by similar arguments; if a $(2q-1)$-knot in a rational sphere bounding a parallelizable rational ball admits a generalized Seifert surface of complexity c, then it is concordant to a knot in a rational sphere bounding a parallelizable rational ball, which admits a $(q-1)$-connected generalized Seifert surface of complexity c. In addition, they are concordant via a concordance of complexity $\leq c$.

The following is a weaker version of our slicing theorem for primitive simple knots:

PROPOSITION 4.10. *Suppose $q > 1$ and K is a primitive $(2q-1)$-knot in a rational sphere Σ bounding a $(q-1)$-connected parallelizable rational ball Δ. If K admits a $(q-1)$-connected Seifert surface F with a metabolic Seifert matrix, then there is a rational $2q$-disk in Δ bounded by K.*

PROOF. Since F is $(q-1)$-connected, $H_q(F)$ is free of even rank, say $2g$, and we can view $H_q(F)$ as a subgroup of
$$H_q(F; \mathbb{Q}) = H_q(F) \otimes \mathbb{Q}.$$
Let $H \subset H_q(F; \mathbb{Q})$ be a metabolizer of the Seifert pairing
$$S \colon H_q(F; \mathbb{Q}) \times H_q(F; \mathbb{Q}) \longrightarrow \mathbb{Q}.$$
Then it can be checked that $H_0 = H \cap H_q(F)$ is a rank g summand of $H_q(F)$. Choose a basis $\{x_i\}$ of H_0 which extends to a basis of $H_q(F)$. Let $r = |\pi_q(\Delta)| = |H_q(\Delta)|$, which is finite.

We claim that the classes rx_i can be represented by disjoint embedded q-spheres α_i in F. For $q > 2$, the claim follows from the Whitney trick since the intersection number of rx_i and rx_j in F is given by $S(rx_i, rx_j) - \epsilon S(rx_j, rx_i) = 0$. For $q = 2$, we again appeal to the arguments in [**33**]. As in the proof of Proposition 4.8, we may assume that F' is obtained by attaching $2g$ 2-handles to a 4-ball along a split union of g Hopf links contained in the boundary of the 4-ball, and furthermore, we may assume that x_i is represented by the core of the 2-handle attached along the first component of the i-th Hopf link by the arguments in [**33**, p. 236]. Now desired spheres α_i are obtained in a similar way as the construction of the complex Y in the proof of Proposition 4.8; α_i is the union of the surface $C_i^1 \subset B_4$ illustrated in Figure 4 and r parallel copies of the core of the i-th 2-handle.

By our choice of r, there are immersed disjoint disks δ_i in Δ bounded by α_i. By Whitney trick again, we may assume that the δ_i are disjoint embedded disks since the intersection number of δ_i and δ_j is given by $S(rx_i, rx_j) = 0$. Appealing to Lemma 4.7 (2), we can do surgery on F using the δ_i, since the self-linking of α_i in Σ is zero. It is easily seen that the resulting submanifold in Δ is a rational disk, which is bounded by K. □

REMARK 4.11. The rational disk constructed in the above proof is $(q-1)$-connected and parallelizable, and has a trivial normal bundle, since F is $(q-1)$-connected and the trace of the ambient surgery in the above proof is a parallelizable two-sided codimension one submanifold in Δ.

Now we are ready to prove our slicing theorem in higher odd dimensions.

THEOREM 4.12. *Suppose $q > 1$ and K is a rational $(2q-1)$-knot in a rational sphere Σ bounding a parallelizable rational ball. If K has a Seifert matrix A such that $i_r A$ is metabolic for some $r > 0$, then K is a rational slice knot, i.e., K bounds an honest $2q$-disk in a rational ball bounded by Σ.*

PROOF. Suppose F is a generalized Seifert surface for K on which the Seifert matrix A is defined. We may assume that F has no closed component by piping (this does not change the Seifert matrix for $q > 1$). Furthermore we may assume that A is metabolic (i.e. $r = 1$) by replacing F by r parallel copies of F.

Suppose F has complexity c, that is, ∂F consists of c parallel copies of K. Since F has no closed component, $\Sigma - F$ is connected. We join components of ∂F using $(c-1)$ bands whose interiors are disjoint to F. It gives us a primitive knot K_0 which is a band sum of ∂F, together with a Seifert surface F_0 of K_0 which is the union of F and the bands. Note that there is a c-punctured disk C in $\Sigma \times [0,1]$ bounded by $K \times 0 \cup -K_0 \times 1$. By Proposition 4.8, K_0 is concordant to a primitive knot $K_1 \subset \Sigma_1$ which has a $(q-1)$-connected Seifert surface F_1. We denote the concordance by (W_0, C_0); W_0 is a rational homology cobordism between Σ and Σ_1 and $C_0 \cong S^{2q-1} \times [0,1]$. Since there is a concordance of complexity 1 between K_0 and K_1, the Seifert matrix of F_1 is also metabolic. By Proposition 4.10, there is a rational $2q$-disk C_1 in a rational $(2q+2)$-ball Δ_1 such that $\partial(\Delta_1, C_1) = (\Sigma_1, K_1)$. Gluing the above pairs along the boundaries, we construct a pair

$$(\Delta, D) = (\Sigma \times [0,1], C) \underset{(\Sigma \times 1, K_0 \times 1)}{\cup} (W_0, C_0) \underset{(\Sigma_1, K_1)}{\cup} (\Delta_1, C_1)$$

of a rational $(2q+2)$-disk Δ and a c-punctured rational $2q$-sphere D which is bounded by $(\Sigma, \partial F)$.

Denote by V the manifold obtained by attaching a $2q$-handle $D^{2q} \times D^2$ to Δ along K. We will construct a slice disk complement U for K by killing $H_{2q}(V;\mathbb{Q}) \cong \mathbb{Q}$ generated by the $2q$-handle. Attaching c parallel copies of the core of the $2q$-handle to D, we obtain a rational $2q$-sphere S in V. To kill $H_{2q}(V;\mathbb{Q})$, we perform "surgery" on V along S as follows (indeed this kills r times the generator of $H_{2q}(V;\mathbb{Z})$ represented by the $2q$-handle). Since C_1 is parallelizable, S is q-parallelizable. If q is odd, we may assume that S has vanishing Arf invariant by replacing the original generalized Seifert surface F with the union of two parallel copies of F at the beginning and applying the above arguments (this gives $S \# S$ instead of S). So, by Lemma 4.5, there is a rational $(2q+1)$-ball B bounded by S. The normal bundle of S in V is trivial, since the obstruction lives in

$$H^2(S; \pi_1(SO_2)) \cong H^2(C_1; \pi_1(SO_2))$$

and C_1 has trivial normal bundle. Identifying a tubular neighborhood of S in V with $S \times D^2$, we remove $\text{int}(S \times D^2)$ from V and fill it in with $B \times S^1$ along the boundary to obtain

$$U = (V - \text{int}(S \times D^2)) \underset{S \times S^1}{\cup} (B \times S^1).$$

Note that ∂U is the surgery manifold of K, and $\tilde{H}_*(U; \mathbb{Q})$ vanishes except $H_1(U; \mathbb{Q}) = \mathbb{Q}$ which is generated by the meridian of K. Attaching a 2-handle $D^2 \times D^{2q}$ to U along the meridian, we obtain a rational ball and Σ is recovered as its boundary. The cocore $0 \times D^{2q}$ of the 2-handle is an honest disk bounded by K. This completes the proof. □

As consequences of Theorem 4.1 and Theorem 4.12, Theorem 1.3 (2), (3), and (4) follow.

4.2.2. Slicing even dimensional rational knots. Using similar techniques, we prove the following slicing theorem in even dimensions.

THEOREM 4.13. *Suppose K is a $2q$-knot in a parallelizable rational $(2q+2)$-sphere Σ. If q is odd, or q is even and Σ has vanishing Arf invariant, then K is a rational slice knot.*

PROOF. By Lemma 4.5, there is a q-connected rational $(2q+3)$-ball Δ bounded by Σ. Choose a generalized Seifert surface F for K. As done in the proof of Proposition 4.8, we will do ambient surgery on F in Δ, to make it q-connected. Since F is parallelizable and $(2q+1)$-dimensional, there is a collection of disjoint spheres of dimension $\leq q$ in the interior of F such that (abstract) surgery along those spheres gives rise to a q-connected manifold. Since Δ is $(2q+3)$-dimensional and q-connected, we can find disjoint disks of dimension $\leq q+1$ in Δ which are bounded by the above spheres. Appealing to Lemma 4.7, we can do the desired ambient surgery using these disks. This gives us an honest sphere with punctures which is properly embedded in Δ and bounded by parallel copies of K. Now the argument of the last part of the proof of Theorem 4.12 can be used to show that K is a rational slice knot. □

Theorem 1.3 (1) is a corollary of Theorem 4.13.

4.3. Rational and integral concordance

In this section we study the natural homomorphism of the integral knot concordance group $\mathcal{C}_n^{\mathbb{Z}}$ into the rational knot concordance group $\mathcal{C}_n$. Since $\mathcal{C}_n^{\mathbb{Z}} = 0$ for even n, we assume that n is odd, say $n = 2q - 1$, throughout this section. Since the image of $\mathcal{C}_n^{\mathbb{Z}} \to \mathcal{C}_n$ is contained in the subgroup $b\mathcal{C}_n$, we will consider the induced homomorphism $\mathcal{C}_n^{\mathbb{Z}} \to b\mathcal{C}_n$. Note that for $n > 1$ there is a commutative diagram

$$\begin{array}{ccc} \mathcal{C}_n^{\mathbb{Z}} & \longrightarrow & b\mathcal{C}_n \\ \downarrow & & \downarrow \\ G_{n,1} = G_n & \xrightarrow{\phi_1} & \mathcal{G}_n = \varinjlim_c G_{n,c} \end{array}$$

where the vertical homomorphisms are injective.

4.3.1. Kernel of $\mathcal{C}_n^{\mathbb{Z}} \to b\mathcal{C}_n$. It has already been known that $\mathcal{C}_n^{\mathbb{Z}} \to b\mathcal{C}_n$ is not injective. In fact, in Example 3.17, we described a Seifert matrix A such that for $n = 4k + 1 > 1$, the concordance class of any n-knot in S^{n+2} with Seifert matrix A is a nontrivial order two element in the kernel of $\mathcal{C}_n^{\mathbb{Z}} \to b\mathcal{C}_n$. We generalize Example 3.17 as follows:

THEOREM 4.14. *For any odd $n > 1$, the kernel of $\mathcal{C}_n^{\mathbb{Z}} \to b\mathcal{C}_n$ contains a subgroup isomorphic to $(\mathbb{Z}/2)^\infty$.*

In the proof of Theorem 4.14, we need the following results of Levine [**33, 32**] on integral Seifert matrices of knots in honest spheres. (Because of different sign conventions, some signs have been changed appropriately)

LEMMA 4.15.
(1) *A polynomial $\Delta(t)$ with integer coefficients is an Alexander polynomial of a $2g \times 2g$ Seifert matrix A of an n-knot in S^{n+2} if and only if $\Delta(t^{-1})t^{2g} = \Delta(t)$, $\Delta(1) = \epsilon^g$, and $\Delta(\epsilon)$ is square.*
(2) *Suppose A is a Seifert matrix of an n-knot in S^{n+2} with Alexander polynomial $\Delta_A(t) = \lambda_1(t)\lambda_2(t)\cdots\lambda_k(t)$ where the $\lambda_i(t)$ are distinct reciprocal irreducible polynomials of degree 2. Then A is of order 2 if and only if A has vanishing signature invariants and, for any $\lambda_i(t)$ and for any prime $p \equiv 3 \mod 4$, the exponent of p in the prime factorization of $\lambda_i(1)\lambda_i(-1)$ is even.*

PROOF OF THEOREM 4.14. First we consider the case q is odd. For a positive integer a, let
$$A = \begin{bmatrix} a & 1 \\ 0 & -a \end{bmatrix}.$$
A is a Seifert matrix of an n-knot K in S^{n+2} by Lemma 4.15. We claim that K is in the kernel of $\mathcal{C}_n^{\mathbb{Z}} \to b\mathcal{C}_n$. It suffices to show that $\phi_1[A] = 0$ in $\mathcal{G}_n$. Since
$$\Delta_A(t) = -a^2 t^2 + (2a^2 + 1)t - a^2,$$
the reparametrization formula gives us
$$\Delta_{i_2 A}(t) = \Delta_A(t^2) = -a^2 t^4 + (2a^2 + 1)t^2 - a^2 = -(at^2 + t - a)(at^2 - t - a).$$
Since both irreducible factors are not reciprocal, $[i_2 A] = 0$ in $\mathcal{G}_n$ by Proposition 3.6. This shows the claim.

Now we show that K has order two in $\mathcal{C}_n^{\mathbb{Z}}$. It suffices to show that $[A]$ is of order two in $\mathcal{G}_n$. Since $\Delta_A(t)$ is irreducible, $[A]$ is nontrivial in $\mathcal{G}_n$ (if one wants, the invariant $e_z(A)$ can be used). Note that A is a Seifert matrix of a 1-knot K_a in S^3 which is illustrated in Figure 5. Since K_a is amphicairal (i.e., K_a is isotopic to $-K_a$), $-[A] = [A]$ in $\mathcal{G}_n$.

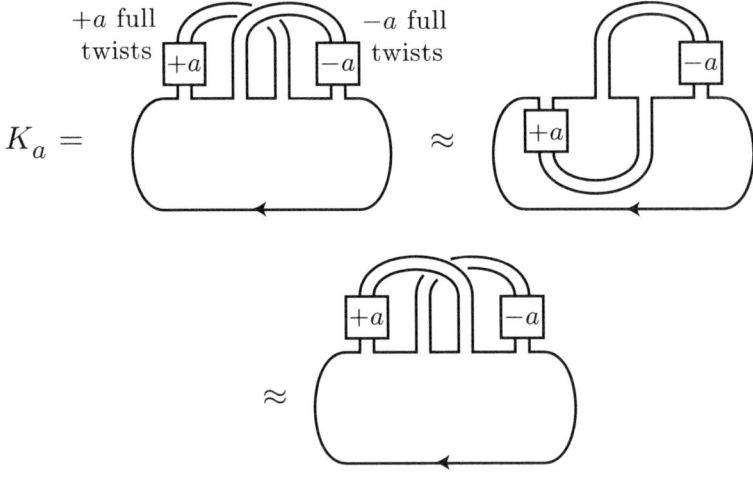

FIGURE 5

Furthermore, different values of a give us different matrices A which are independent in $\mathcal{G}_n$ since they have relatively prime Alexander polynomials (one may use $e_z(A)$ again). Therefore the associated knots K are also independent. This completes the proof for odd q. For later use, we observe the following fact: by Lemma 4.15 (2), the exponent of p in the prime factorization of $-(4a^2 + 1) = \Delta_A(1)\Delta_A(-1)$ is even for any prime $p \equiv 3$ mod 4.

Now we consider the case q is even. Let
$$\Delta(t) = (t^2 - 3t + 1)(a^2 t^2 - (2a^2 + 1)t + a^2),$$

4.3. RATIONAL AND INTEGRAL CONCORDANCE

where a is a positive integer such that $\Delta(-1) = 5(4a^2 + 1)$ is square. Then by Lemma 4.15 (1), $\Delta(t)$ is the Alexander polynomial of a Seifert matrix A of an n-knot K in S^{n+2}. As before, by observing the factorization of $\Delta_{i_2 A}(t)$, $[i_2 A] = 0$ in G_n and thus $[A]$ is in the kernel of $G_n \to \mathcal{G}_n$. It follows that K is in the kernel of $\mathcal{C}_n^{\mathbb{Z}} \to b\mathcal{C}_n$.

Note that $\Delta(t)$ has two irreducible factors $f(t) = t^2 - 3t + 1$ and $g(t) = a^2 t^2 - (2a^2 + 1)t + a^2$. By the observation above, for any prime $p \equiv 3 \mod 4$, the exponents of p in the factorization of $f(1)f(-1) = -1$ and $g(1)g(-1) = -(4a^2 + 1)$ are always even. Therefore, by Lemma 4.15 (2), A has order two in G_n. It follows that K has order two in $\mathcal{C}_n^{\mathbb{Z}}$.

As before, different values of a gives us different Seifert matrices A which are independent in G_n. Therefore, to complete the proof, it suffices to show that there are infinitely many a such that $5(4a^2 + 1)$ is square. For this purpose, we consider a Diophantine equation

$$x^2 - 5y^2 = -1$$

which is a specific form of Pell's equation. It is known that there are infinitely many solutions (x, y) of this equation. A concrete description is as follows. Let $(x_0, y_0) = (1, 0)$, $(x_1, y_1) = (2, 1)$, and

$$x_{n+2} = 4x_{n+1} + x_n$$
$$y_{n+2} = 4y_{n+1} + y_n$$

for $n \geq 0$. Then it can be shown that $x_n^2 - 5y_n^2 = (-1)^n$ by an induction. In particular, (x_{2n+1}, y_{2n+1}) is a solution of our Diophantine equation. These solutions are different since $\{x_i\}$ is increasing. Since x_{2n+1} is even and $5(x_{2n+1}^2 + 1) = (5y_{2n+1})^2$, the integer $a = x_{2n+1}/2$ has the desired property. This completes the proof for even q. □

For $n = 1$, the above arguments do not work. However, Cochran proved that the kernel of $\mathcal{C}_1^{\mathbb{Z}} \to b\mathcal{C}_1$ is nontrivial. In fact he showed that the figure eight knot, which has order two in $\mathcal{C}_n^{\mathbb{Z}}$, is a rational slice knot using a Kirby calculus argument similar to that of Fintushel and Stern [17]. Generalizing his arguments, we prove the following result.

THEOREM 4.16. *The kernel of $\mathcal{C}_1^{\mathbb{Z}} \to b\mathcal{C}_1$ contains a subgroup isomorphic to $(\mathbb{Z}/2)^\infty$.*

PROOF. We will show that the knot K_a in S^3 illustrated in Figure 5 is a rational slice knot. Since the concordance classes of the K_a have order two in $\mathcal{C}_1^{\mathbb{Z}}$ and are independent, as we observed in the proof of Theorem 4.14, it proves the desired conclusion.

Let M be the 3-manifold obtained by null-framed surgery on S^3 along K_a. In a similar way as [17], we will construct a rational homology cobordism between M and $S^2 \times S^1$. First, starting with $M \times [0, 1]$, we construct a cobordism W_1 between M and another manifold M' as illustrated by the Kirby diagrams in Figure 6. W_1 is obtained by attaching a 1-handle and a 2-handle, and it can be seen that the 2-handle kills the generator introduced by the 1-handle (over the rationals). Thus W_1 is a rational homology cobordism.

On the other hand, Figure 7 illustrates that the underlying link of the Kirby diagram describing M' is concordant to the link L shown in the last diagram in Figure 7. Figure 7 can be viewed as a "movie" illustration of a concordance in $S^3 \times [0,1]$; it illustrates the intersection of the concordance with the level spheres $S^3 \times t$. From this it follows that there is a $\mathbb{Z}$-homology cobordism between M' and the result of surgery M'' along L with respect to framings 2, 0, and -2. It is easily seen that $M'' \cong S^2 \times S^1$, and thus $W = W_1 \cup_{M'} W_2$ is a rational homology cobordism between M and $S^2 \times S^1$.

Let $V = W \cup_{S^2 \times S^1} D^3 \times S^1$. V is bounded by M and has the rational homology of S^1, where $H_1(V;\mathbb{Q})$ is generated by the meridian of K_a. Attaching a 2-handle to V along the meridian of K_a, we obtain a rational ball bounded by S^3. The cocore of the 2-handle is an honest 2-disk bounded by K_a. This shows that K_a is a rational slice knot. □

4.3.2. Cokernel of $\mathcal{C}_n^{\mathbb{Z}} \to b\mathcal{C}_n$. We will investigate the structure of the cokernel of $G_n \to \mathcal{G}_n$ and then pull it back along

$$\operatorname{Coker}\{\mathcal{C}_n^{\mathbb{Z}} \longrightarrow b\mathcal{C}_n\} \longrightarrow \operatorname{Coker}\{G_n \longrightarrow \mathcal{G}_n\}.$$

In [12] and [7], some periodicity of the signature invariant of $\mathcal{G}_n$ was used to investigate $\operatorname{Coker}\{G_n \to \mathcal{G}_n\}$. We generalize it for our invariants of $\mathcal{G}_n$. Recall that for $\mathcal{A} \in \mathcal{G}_n$, $s_\alpha(\mathcal{A})$, $e_\alpha(\mathcal{A})$, and $s_\alpha(\mathcal{A})$ denote the "α-coordinates" of the invariants $s(\mathcal{A})$, $e(\mathcal{A})$, and $d(\mathcal{A})$, respectively. (See the last part of Section 3.2.)

THEOREM 4.17. *If $\mathcal{A} \in \mathcal{G}_n$ is contained in the image of $\phi_c \colon G_{n,c} \to \mathcal{G}_n$, then for any $\alpha = (\alpha_i)$ and $\beta = (\beta_i) \in P$ such that $\alpha_c = \beta_c$, the followings hold:*

(1) $s_\alpha(\mathcal{A}) = s_\beta(\mathcal{A})$ *provided* $\alpha, \beta \in P_0$.
(2) $e_\alpha(\mathcal{A}) = e_\beta(\mathcal{A})$.
(3) $d_\alpha(\mathcal{A}) = d_\beta(\mathcal{A})$.

PROOF. Suppose that $[A] \in G_n = G_{n,c}$ is sent to $\mathcal{A} \in \mathcal{G}_n$ via ϕ_c. Since α_c and β_c are the same complex numbers, we have $e_{\alpha_c}[A] = e_{\beta_c}[A]$. (Recall that $e_z[A]$ is the modulo 2 residue class of the rank of the α_c-primary part of $[A] \in G_n$.) By our definition,

$$e_\alpha(\mathcal{A}) = e_{\alpha_c}[A] = e_{\beta_c}[A] = e_\beta(\mathcal{A}).$$

The same argument works for the invariants s and d. □

Now we apply Theorem 4.17 to study the structure of the torsion part of $\operatorname{Coker}\{G_n \to \mathcal{G}_n\}$. Recall that in Corollary 3.40 we constructed a summand H of (the torsion part of) $\mathcal{G}_n$ isomorphic to $(\mathbb{Z}/2)^\infty \oplus (\mathbb{Z}/4)^\infty$. H is generated by order 2 elements of the form $\phi_c[A_i]$ and order 4 elements of the form

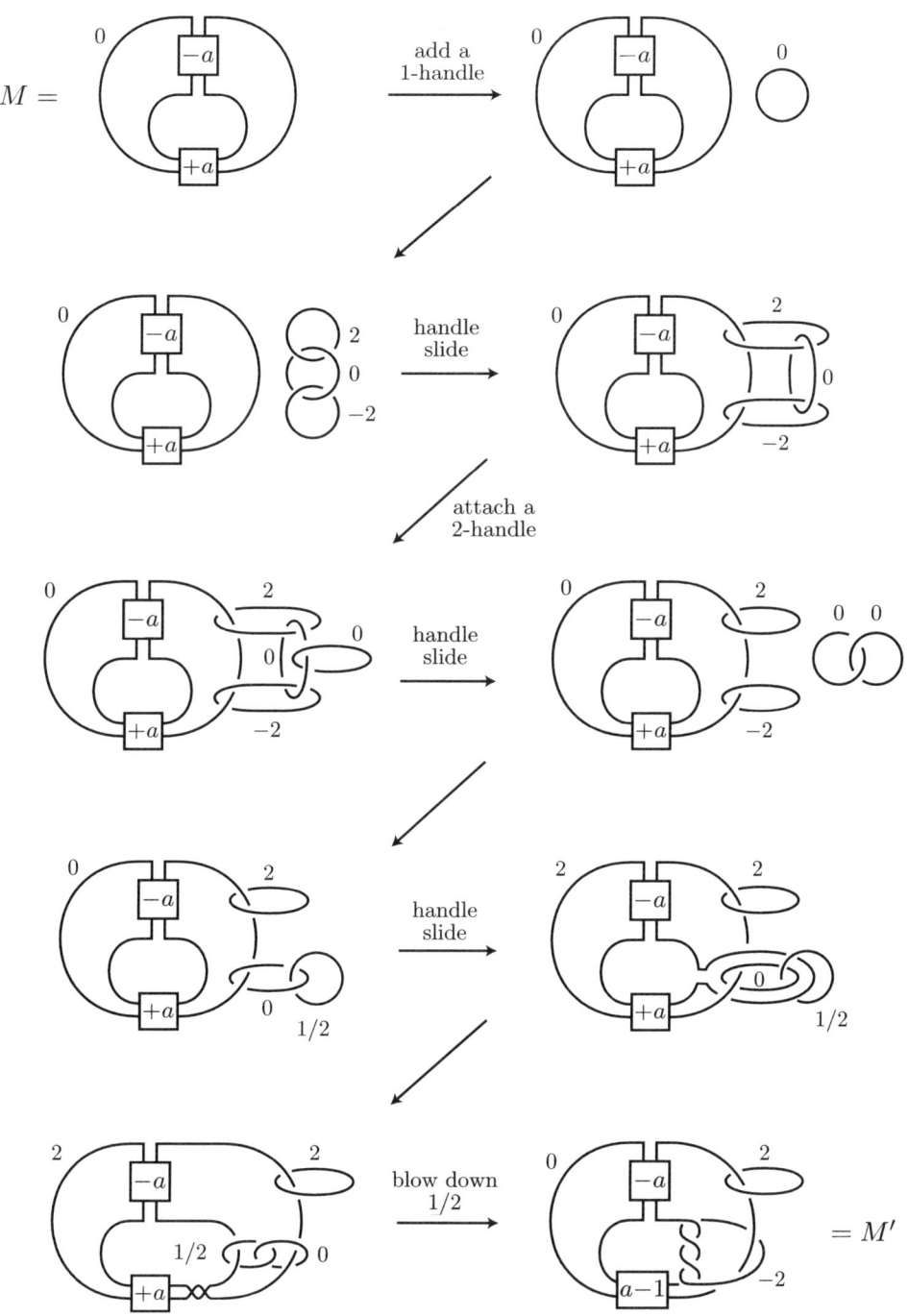

FIGURE 6

$\phi_c[B_i]$, where $[A_i], [B_i] \in G_n$ and $c > 0$ is a positive integer. Henceforth we

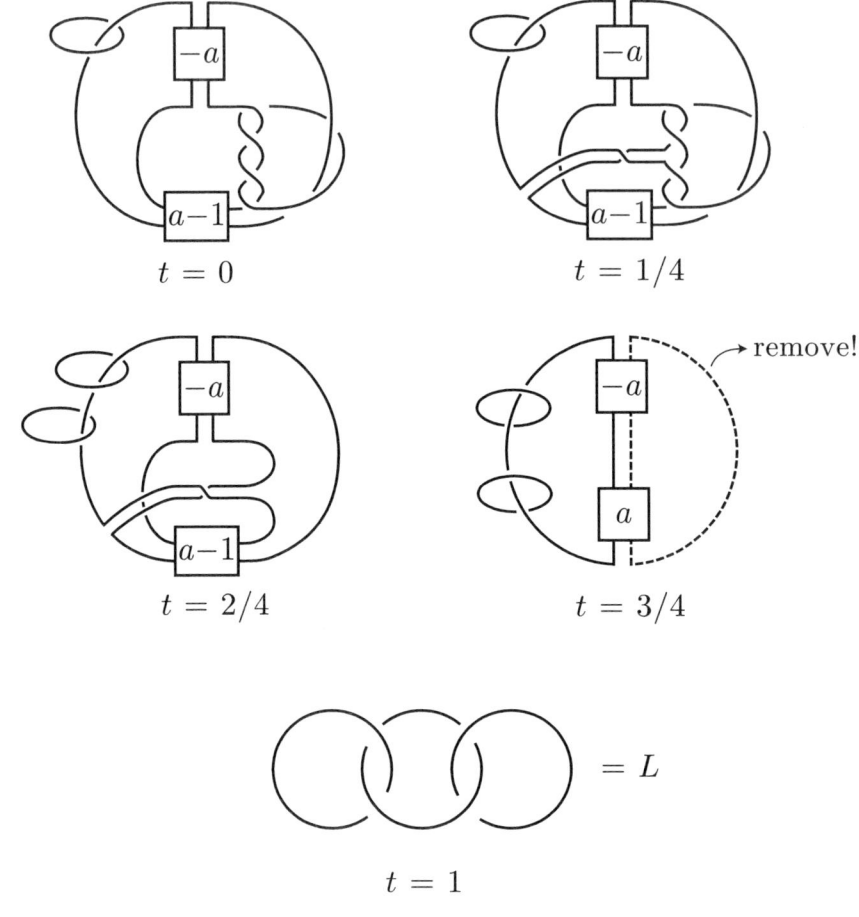

FIGURE 7

fix $c = 2$ and consider the subgroup
$$H \subset \operatorname{Im}\{\phi_2\colon G_n \longrightarrow \mathcal{G}_n\} \subset \mathcal{G}_n$$
generated by $\mathcal{A}_i = \phi_2[A_i]$ and $\mathcal{B}_i = \phi_2[B_i]$.

PROPOSITION 4.18. $H \cap \operatorname{Im}\{\phi_1\colon G_n \to \mathcal{G}_n\} = \{0\}$.

PROOF. As in the proof of Theorem 3.38, choose zeros $z_i, w_i \neq -1$ of $\Delta_{A_i}(t)$, $\Delta_{B_i}(t)$ and choose $\alpha_i, \beta_i \in P$ such that $(\alpha_i)_2 = z_i$, $(\beta_i)_2 = w_i$. In addition, by Remark 3.21, there are $\alpha_i', \beta_i' \in P$ such that $(\alpha_i')_2 = -z_i$, $(\beta_i')_2 = -w_i$. Since the $(-z)$-primary parts of A_i, B_i are trivial for any $z \in \{z_i, w_i\}$, we have the following property, in addition to the properties (1) and (2) in the proof of Theorem 3.38:

(3) For $\alpha \in \{\alpha_i', \beta_i'\}$ and $A \in \{A_i, B_i\}$, $e_\alpha(A)$ and $d_\alpha(A)$ are trivial.

Suppose that $\mathcal{A} = \sum a_i \mathcal{A}_i + \sum b_i \mathcal{B}_i$ is contained in $\operatorname{Im}\{\phi_1\}$, where a_i, b_i are integers. Observe that
$$(\alpha_i)_1 = z_i^2 = (-z_i)^2 = (\alpha_i')_1,$$

and similarly $(\beta_i)_1 = (\beta'_i)_1$. Thus by Theorem 4.17, we have $e_{\alpha_i}(\mathcal{A}) = e_{\alpha'_i}(\mathcal{A})$ and $e_{\beta_i}(\mathcal{A}) = e_{\beta'_i}(\mathcal{A})$. Taking e_{α_i} and $e_{\alpha'_i}$ of $\sum a_i \mathcal{A}_i + \sum b_i \mathcal{B}_i$, it follows that each a_i is even, from the properties (1) and (3). Considering $e_{\beta_i}(\mathcal{A})$ and $e_{\beta'_i}(\mathcal{A})$ in a similar way, we can see that each b_i is even. Letting $b_i = 2b'_i$, $\mathcal{A} = \sum b'_i (2\mathcal{B}_i)$. Now since $d_{\beta_i}(\mathcal{A}) = d_{\beta'_i}(\mathcal{A})$, it follows that b'_i is even from the properties (2) and (3). This shows that $\mathcal{A} = 0$. □

In [7, Theorem 1.3], it was proved that $\operatorname{Coker} \phi_1$ contains $\mathbb{Z}^\infty$. Indeed its proof shows, using signature invariants, that there is a subgroup $H' \cong \mathbb{Z}^\infty$ in $\mathcal{G}_n$ such that $H' \cap (\operatorname{Im} \phi_1 + T) = \{0\}$, where T is the torsion subgroup of $\mathcal{G}_n$. From Proposition 4.18 and the fact that H is a summand of T, it follows that $H' \oplus H$ is a summand of $\mathcal{G}_n$ such that

$$(H' \oplus H) \cap \operatorname{Im} \phi_1 = \{0\}.$$

Therefore we have the following consequence:

COROLLARY 4.19. *Coker* ϕ_1 *has a direct summand isomorphic to* $\mathbb{Z}^\infty \oplus (\mathbb{Z}/2)^\infty \oplus (\mathbb{Z}/4)^\infty$.

Note that, for odd $n > 3$, $\operatorname{Coker}\{\mathcal{C}_n^{\mathbb{Z}} \to b\mathcal{C}_n\} \cong \mathcal{G}_n/S$ where S is the image of the composition

$$\mathcal{C}_n^{\mathbb{Z}} \longrightarrow \mathcal{G}_n \xrightarrow{\phi_1} \mathcal{G}_n.$$

Since $S \subset \operatorname{Im} \phi_1$, $(H' \oplus H) \cap S = \{0\}$. Thus the second conclusion of Theorem 1.4 follows: $\operatorname{Coker}\{\mathcal{C}_n^{\mathbb{Z}} \to b\mathcal{C}_n\}$ contains a summand isomorphic to $\mathbb{Z}^\infty \oplus (\mathbb{Z}/2)^\infty \oplus (\mathbb{Z}/4)^\infty$. For $n = 3$, by replacing $\mathcal{G}_n$ with its index two subgroup $\operatorname{Im}\{\mathcal{C}_n \to \mathcal{G}_n\}$, the same argument works.

REMARK 4.20. On the same lines as Remark 3.42, the results in this section can be rephrased in terms of surgery obstruction Γ-groups: for odd n, the kernel and cokernel of the homomorphism

$$\Gamma_{n+3}\begin{pmatrix} \mathbb{Z}[\mathbb{Z}] \xrightarrow{\text{id}} \mathbb{Z}[\mathbb{Z}] \\ \text{id}\downarrow \quad \downarrow \varepsilon \\ \mathbb{Z}[\mathbb{Z}] \xrightarrow{\varepsilon} \mathbb{Z} \end{pmatrix} \longrightarrow \Gamma_{n+3}\begin{pmatrix} \mathbb{Q}[\mathbb{Q}] \xrightarrow{\text{id}} \mathbb{Q}[\mathbb{Q}] \\ \text{id}\downarrow \quad \downarrow \varepsilon \\ \mathbb{Q}[\mathbb{Q}] \xrightarrow{\varepsilon} \mathbb{Q} \end{pmatrix}$$

have a subgroup isomorphic to $(\mathbb{Z}/2)^\infty$ and a summand isomorphic to $\mathbb{Z}^\infty \oplus (\mathbb{Z}/2)^\infty \oplus (\mathbb{Z}/4)^\infty$, respectively.

4.4. Subrings of rationals

We remark that, for any subring R of $\mathbb{Q}$, most of our higher-dimensional arguments can be applied to knots in R-homology spheres. Let S be a set consisting of primes, I be the set of positive integers which are coprime to all $p \in S$, and R be the subring of $\mathbb{Q}$ generated by $\{1/c \in \mathbb{Q} \mid c \in I\}$. We can define R-concordance of n-knots in R-homology $(n+2)$-spheres and the R-concordance group $\mathcal{C}_n^R$ of such knots in an obvious way. The geometric arguments in Sections 4.1 and 4.2 also work in this case. For odd n, instead of our $G_{n,i}$ and $\mathcal{G}_n$, we need to consider the algebraic concordance group $G_{n,i}^R$ of Seifert matrices over R and their limit

$$\mathcal{G}_n^R = \varinjlim_{i \in I} G_{n,i}^R.$$

Then a homomorphism $\mathcal{C}_n^R \to \mathcal{G}_n^R$ is defined, by appealing to the property that the complexity of a knot in an R-homology sphere is coprime to all $p \in S$, which follows from the Alexander duality with R-coefficients. Most proofs carry over R, although some statements need to be modified a little; e.g., in case of Theorem 3.22 (3), "nonzero square in $\mathbb{Q}$" should be read as "unit square in R".

In particular, the structure of $\mathcal{G}_n^R$ can be calculated in the same way as Sections 3.2, 3.3, and 3.5. If $2 \notin S$, we can obtain complete invariants of $\mathcal{G}_n^R$, using parameter sets

$$P^R = \{(\alpha_i)_{i \in I} \mid (\alpha_i r)^r = \alpha^r \text{ for } r \in I\},$$
$$P_0^R = \{(\alpha_i) \in P^R \mid |\alpha_i| = 1\},$$

and in this case our algebraic construction of torsion elements also works in $\mathcal{G}_n^R$.

If $2 \in S$, the situation is so simpler that we do not need to use the full power of our algebraic results. Indeed in this case the morphisms $G_{n,i}^R \to G_{n,ri}^R$ defining the limit $\mathcal{G}_n^R$ are all injective (e.g., see [**12**, Proposition 2.1]). Since all the elements in $G_{n,i}^R$ survive in $\mathcal{G}_n^R$, the existence of torsion elements in $\mathcal{G}_n^R$ is immediate. From this it follows that all the analogues of the theorems in the introduction (Chapter 1) hold for any R. Also, the analogues of the results on Γ-groups discussed in Remarks 3.42 and 4.20 hold.

CHAPTER 5

Rational knots in dimension three

5.1. Rational (0)- and (0.5)-solvability

Let G be a group. For two elements a and b in G, the commutator of a and b is defined by $[a,b] = aba^{-1}b^{-1}$. For two subgroups A and B in G, we define $[A,B]$ to be the subgroup generated by $\{[a,b] \mid a \in A \text{ and } b \in B\}$. The *n-th derived subgroup* $G^{(n)}$ is defined inductively by

$$G^{(0)} = G, \quad G^{(n+1)} = [G^{(n)}, G^{(n)}].$$

To kill torsion elements in quotients of G by derived subgroups, we consider the *rational derived subgroup* $G_\mathbb{Q}^{(n)}$ as in [11] and [19]:

DEFINITION 5.1. The *n-th rational derived subgroup* $G_\mathbb{Q}^{(n)}$ is defined by

$$G_\mathbb{Q}^{(0)} = G, \quad G_\mathbb{Q}^{(n+1)} = \{g \in G_\mathbb{Q}^{(n)} \mid g^r \in [G_\mathbb{Q}^{(n)}, G_\mathbb{Q}^{(n)}] \text{ for some } r \neq 0\}.$$

It is known that $G_\mathbb{Q}^{(n)}$ is a normal subgroup of G [11, 19].

For a 4-manifold W with fundamental group G, there is an intersection form on the homology of W with $\mathbb{Q}[G/G_\mathbb{Q}^{(n)}]$-coefficients:

$$\lambda_n \colon H_2(W; \mathbb{Q}[G/G_\mathbb{Q}^{(n)}]) \times H_2(W; \mathbb{Q}[G/G_\mathbb{Q}^{(n)}]) \longrightarrow \mathbb{Q}[G/G_\mathbb{Q}^{(n)}].$$

Denote by $X^{(n)}$ the regular cover of a CW-complex X associated to the normal subgroup $\pi_1(X)_\mathbb{Q}^{(n)}$ in $\pi_1(X)$. The homology module $H_2(W; \mathbb{Q}[G/G_\mathbb{Q}^{(n)}])$ is identified with the rational homology $H_2(W^{(n)}; \mathbb{Q})$ of the cover $W^{(n)}$.

Now we are ready to define the rational solvability of 3-manifolds and 1-dimensional knots following [13].

DEFINITION 5.2. For a closed 3-manifold M and a nonnegative integer n, a 4-manifold W is called a *rational (n)-solution* of M if

(1) $\partial W = M$,
(2) $H_1(M; \mathbb{Q}) \to H_1(W; \mathbb{Q})$ is an isomorphism, and
(3) there exist elements

$$v_1, \ldots, v_m, u_1, \ldots, u_m \in H_2(W^{(n)}; \mathbb{Q})$$

such that $\lambda_n(u_i, u_j) = 0$, $\lambda_n(u_i, v_j) = \delta_{ij}$ (the Kronecker delta), and the images of the u_i, v_i in $H_2(W; \mathbb{Q})$ under the covering map form a basis of $H_2(W; \mathbb{Q})$.

We call W a *rational $(n.5)$-solution* of M if, in addition to (1), (2), and (3) above,

(4) there exist elements
$$u'_1, \ldots, u'_m \in H_1(W^{(n+1)}; \mathbb{Q})$$
such that $\lambda_{n+1}(u'_i, u'_j) = 0$ and the above u_i is the image of u'_i under the covering map.

If there is a (h)-solution W, then M is called *rationally (h)-solvable*.

REMARK 5.3. Definition 5.2 is slightly different from the original definition of Cochran, Orr, and Teichner [**13**, Definition 4.1] which uses the ordinary derived series $G^{(n)}$ instead of the rational derived series $G_{\mathbb{Q}}^{(n)}$. In fact, it turns out that our definition is a more accurate description of the geometric property which is detected by solvable poly-torsion-free-abelian (PTFA) coefficient systems. A rational (h)-solution in the sense of [**13**, Definition 4.1] is a rational (h)-solution in the sense of Definition 5.2. All the results in [**13**] about rational solvability hold when the original definition is replaced by Definition 5.2.

Henceforth we consider only 1-dimensional (rational) knots. In order to apply the notion of rational solvability to knots, we consider the surgery manifold obtained by filling the exterior with a solid torus. For this we need a fixed choice of a framing. In case of integral knots, the zero-linking framing is an obvious choice. However, a rational knot might not allow such a canonical framing; every pushoff of a given knot may have nontrivial linking number with the knot, since the linking number is rational-valued. Hence we are naturally led to consider knots admitting a longitude with linking number zero. It is equivalent to the condition that the knot has vanishing $(\mathbb{Q}/\mathbb{Z})$-valued self-linking, or that there exists a generalized Seifert surface, by Theorem 2.6. Note that this is a necessary condition for a knot to be a rational slice knot. For such knots, we call the framing induced by a generalized Seifert surface the *zero-framing* and call the result of surgery along this framing the *zero-surgery manifold*.

DEFINITION 5.4. A rational knot K with vanishing $(\mathbb{Q}/\mathbb{Z})$-valued self-linking is called *rationally (h)-solvable* if the zero-surgery manifold M of K is rationally (h)-solvable.

The subgroup of the classes of rationally (h)-solvable knots in $s\mathcal{C}_1 \subset \mathcal{C}_1$ is denoted by $\mathcal{F}_{(h)}^{\mathbb{Q}}$.

REMARK 5.5. In case of a knot in S^3, the ordinary derived subgroup and the rational derived subgroup of the fundamental group of the zero-surgery manifold are equal, so that the definitions of rational solvability in this paper and [**13**] are equivalent.

In Definition 2.8, the complexity of a knot K was defined in terms of the meridian in the integral homology of the knot exterior. It is easily seen that the complexity can also be defined by considering the surgery manifold M, instead of the exterior: $H_1(M;\mathbb{Q}) = \mathbb{Q}$ is generated by the meridian of K by the Alexander duality, and hence $H_1(M;\mathbb{Z})/\text{torsion}$ is isomorphic to $\mathbb{Z}$. The meridian of K generates a nontrivial subgroup in $\mathbb{Z}$, and its positive generator is the complexity of K. This is equivalent to our previous definition.

The complexity of a rational solution W of M is defined in a similar way. Since $H_1(W;\mathbb{Q}) \cong H_1(M;\mathbb{Q}) = \mathbb{Q}$, $H_1(W;\mathbb{Z})/\text{torsion}$ is isomorphic to $\mathbb{Z}$, and the meridian of K is a nontrivial element in this group.

DEFINITION 5.6. For a rational (h)-solution W of the zero-surgery manifold M of a knot K, the positive generator of the subgroup generated by the meridian of K in $H_1(W;\mathbb{Z})/\text{torsion} \cong \mathbb{Z}$ is called the *complexity* of W.

Since $H_1(M;\mathbb{Z})/\text{torsion} \to H_1(W;\mathbb{Z})/\text{torsion}$ is injective, the complexity of a knot is a divisor of the complexity of its rational solution.

Of course a rational slice knot is rationally (h)-solvable for any h. Indeed a rational solution of a knot can be viewed as an "approximation" of a rational slice disk complement, and in this sense, the rational solvability is a measurement of "how close" a knot is to a rational slice knot.

Similarly, we can think of the rational solvability of a rational 3-sphere as a measurement of the extent of failure to bound a rational 4-ball. Generalizing naively the fact that the ambient space of a rational slice knot must bound a rational 4-ball, one may expect that if a knot is "close" to a rational slice knot, then its ambient space must be "close" to the boundary of a rational 4-ball in the above sense. Indeed it follows from the proposition below, which is a general statement that certain surgery preserves rational solvability.

PROPOSITION 5.7. *Suppose M is a rationally (h)-solvable 3-manifold and $i\colon S^1 \times D^2 \to M$ is an embedding such that $[i(*\times S^1)] = 0$ in $H_1(M - S^1 \times 0;\mathbb{Q})$ and $[i(S^1 \times *)] \neq 0$ in $H_1(M;\mathbb{Q})$, where $* \in S^1$. Then the 3-manifold*

$$N = M - \operatorname{int} i(S^1 \times D^2) \underset{S^1 \times S^1}{\cup} D^2 \times S^1$$

obtained from M by surgery is also rationally (h)-solvable.

PROOF. Suppose W is a rational (h)-solution of M. Let V be the 4-manifold obtained from W by attaching a 2-handle along $i(S^1 \times D^2)$ in such a way that $\partial V = N$. We will show that V is a rational (h)-solution of N.

Let denote $\mu = i(S^1 \times *)$. Then obviously $\pi_1(V) = \pi_1(W)/H$ where H is the normal subgroup generated by μ. Furthermore, by our hypothesis, $H_1(N;\mathbb{Q}) = H_1(M;\mathbb{Q})/\langle\mu\rangle$. Combining this with $H_1(M;\mathbb{Q}) \cong H_1(W;\mathbb{Q})$, it follows that the map $H_1(N;\mathbb{Q}) \to H_1(V;\mathbb{Q})$ induced by the inclusion is an isomorphism.

Next we verify the intersection pairing condition. Let

$$R'_n = \mathbb{Q}[\pi_1(W)/\pi_1(W)^{(n)}_{\mathbb{Q}}],$$
$$R_n = \mathbb{Q}[\pi_1(V)/\pi_1(V)^{(n)}_{\mathbb{Q}}].$$

Since the surjection $\pi_1(W) \to \pi_1(V)$ sends $\pi_1(W)^{(n)}_{\mathbb{Q}}$ into $\pi_1(V)^{(n)}_{\mathbb{Q}}$, it gives rise to a ring homomorphism of R'_n into R_n. Hence we can think of the R_n-coefficient homology $H_*(W; R_n)$ as well as $H_*(W; R'_n)$. Consider the following diagram:

$$H_2(W; R'_n) \twoheadrightarrow H_2(W; R_n) \twoheadrightarrow H_2(V; R_n)$$
$$\downarrow \qquad\qquad \downarrow$$
$$H_2(\mu; \mathbb{Q}) \longrightarrow H_2(W; \mathbb{Q}) \xrightarrow{\alpha} H_2(V; \mathbb{Q}) \longrightarrow H_1(\mu; \mathbb{Q}) \xrightarrow{\beta} H_1(W; \mathbb{Q})$$

The bottom row is exact by a Mayer–Vietoris argument. Since $H_2(\mu; \mathbb{Q}) = 0$ and β is an injection, α is an isomorphism. If $h = n$ is an integer, by the rational (n)-solvability of W, there are elements $u_i, v_j \in H_2(W; R'_n)$ satisfying Definition 5.2. From the naturality of the intersection pairing and that $H_2(W; \mathbb{Q}) \cong H_2(V; \mathbb{Q})$, it follows that the images $\bar{u}_i, \bar{v}_j$ in $H_2(V; R_n)$ also satisfy Definition 5.2. This shows that V is a rational (n)-solution of Σ. Similar argument works for the case $h = n + 0.5$. $\square$

As a consequence, if Σ is a rational 3-sphere and K is a rationally (h)-solvable knot in Σ with vanishing $(\mathbb{Q}/\mathbb{Z})$-valued self-linking, then letting the zero-surgery manifold and Σ play the role of M and N above, respectively, it follows that Σ must be rationally (h)-solvable by Proposition 5.7.

However, by Proposition 5.8 below, it turns out that this application to rational knots is less interesting.

PROPOSITION 5.8. *If a rational 3-sphere Σ is rationally (0)-solvable, then Σ is rationally (h)-solvable for any h.*

PROOF. Suppose W is a rational (0)-solution of Σ. Then $H_1(W; \mathbb{Q}) = H_1(\Sigma; \mathbb{Q})$ vanishes and thus

$$\pi_1(W)/\pi_1(W)^{(1)}_{\mathbb{Q}} = H_1(W; \mathbb{Z})/\text{torsion} = 0.$$

It follows that $\pi_1(W)^{(n)}_{\mathbb{Q}} = \pi_1(W)$ for all n, that is, the cover $W^{(n)}$ is nothing more than W itself. Therefore W is a rational (h)-solution for any h by Definition 5.2. $\square$

REMARK 5.9. From Proposition 5.8, we can also see that any attempt to find further obstructions for a rational 3-sphere to bound a rational 4-ball from PTFA coefficient systems (and associated von Neumann invariants) in a similar way as [**13**] will fail.

5.1. RATIONAL (0)- AND (0.5)-SOLVABILITY

From the viewpoint of knot theory, our main interest is to investigate further obstructions to being rationally (h)-solvable obtained from the complication of knotting, beyond rational solvability of ambient spaces. The following result deals with the cases of $h = 0$ and 0.5.

THEOREM 5.10. *Suppose K is a knot in a rational sphere Σ with vanishing $(\mathbb{Q}/\mathbb{Z})$-valued self-linking. Then*
  (1) *K is rationally (0)-solvable if and only if so is Σ.*
  (2) *K is rationally (0.5)-solvable if and only if Σ is rationally (0)-solvable and there exists a generalized Seifert surface with a metabolic Seifert matrix.*

REMARK 5.11. This result may be compared with the analogues for integral knots discussed in [13]: an integral knot is integrally (0)-solvable if and only if the Arf invariant is zero, and is integrally (0.5)-solvable if and only if its Seifert matrix is metabolic. For integral knots the ambient space condition is unnecessary since S^3 bounds D^4. For rational knots, we have no condition on the Arf invariant; we have no Arf invariant over $\mathbb{Q}$. We note that the Arf invariant condition for integral knots is required since an integral (0)-solution must be a *spin* 4-manifold by definition.

Recall a special case of Theorem 4.1: for any pair of a positive integer c and a rational Seifert matrix A, there is a knot K in a rational 3-sphere Σ bounding a rational 4-ball which has a generalized Seifert surface of complexity c with Seifert matrix A. In particular, this Σ is rationally (0)-solvable, and hence K is rationally (0)-solvable by Theorem 5.10 (1). From this it follows that any element in $\mathcal{G}_1$ is realized by a rationally (0)-solvable knot. Combining this with Theorem 5.10 (2), we obtain Theorem 1.6: $\mathcal{F}^{\mathbb{Q}}_{(0)}/\mathcal{F}^{\mathbb{Q}}_{(0.5)} \cong \mathcal{G}_1$.

PROOF OF THEOREM 5.10. First we prove the if direction. Suppose that Δ is a rational (h)-solution of Σ where $h = 0$ or 0.5. Let V be the manifold obtained from Δ by attaching a 2-handle along the zero-framing of K so that ∂V is the zero-surgery manifold M.

Let F_0 is a generalized Seifert surface in Σ. Pushing the interior of F_0 into the interior of Δ and attaching parallel copies of the core of the 2-handle, we obtain a closed surface F in V. Since F is boundary parallel, there is a canonical framing of the normal bundle of F, and using this, we identify a regular neighborhood of F in V with $F \times D^2$.

Let $X = V - \text{int}(F \times D^2)$. Choose a handlebody R bounded by F, and let
$$W = X \underset{F \times S^1}{\cup} R \times S^1.$$
We will show that W is a rational (h)-solution of M. First, from duality and Mayer–Vietoris arguments it follows that
$$\pi_1(W)/\pi_1(W)^{(1)}_{\mathbb{Q}} \cong H_1(W; \mathbb{Z})/\text{torsion} \cong \mathbb{Z}$$

and it is generated by a meridian of F (not a meridian of K!). Therefore $H_1(M) \to H_1(W)$ is an isomorphism. (Throughout this proof $H_i(-)$ designates $H_i(-;\mathbb{Q})$.)

Now we compute the second homology and the intersection pairing of W. Let $\tilde{W} = W^{(1)}$ be the infinite cyclic cover of W induced by

$$\pi_1(W) \longrightarrow \pi_1(W)/\pi_1(W)^{(1)}_\mathbb{Q} = \mathbb{Z}.$$

We will use a standard cut-paste construction of $\tilde{W}$. Denoting the infinite cyclic cover of X by $\tilde{X}$, we have

$$\tilde{W} = \tilde{X} \underset{F \times \mathbb{R}}{\cup} R \times \mathbb{R}.$$

Recall that F is boundary parallel in V. Thus there is a proper embedding $f: F \times [0,1] \to X$ such that $f(F \times 1) \subset \partial V \subset \partial X$ and $f(F \times 0) \subset \partial X - \partial V$ induces our framing on F in V. Let Y be X cut along $f(F \times [0,1])$. There are inclusions

$$i_+, i_-: F \times [0,1] \longrightarrow \partial Y$$

corresponding to the positive and negative normal directions of $f(F \times [0,1])$ in X, respectively. Then $\tilde{X}$ is given by

$$\tilde{X} = \left(\coprod_{n \in \mathbb{Z}} t^n Y \right) \Big/ i_-(z) \sim t i_+(z) \text{ for } z \in F \times [0,1],$$

where $t^n Y$ is a copy of Y so that t can be viewed as a deck translation in a natural way.

Since $Y \cong V = \Delta \cup (\text{2-handle})$,

$$H_i(Y) = \begin{cases} H_i(\Delta) & \text{for } i \neq 2, \\ H_2(\Delta) \oplus \mathbb{Q} & \text{for } i = 2. \end{cases}$$

By a Mayer–Vietoris argument, there is an exact sequence

$$\cdots \longrightarrow \bigoplus_{n \in \mathbb{Z}} H_2(F) \xrightarrow{\alpha} \bigoplus_{n \in \mathbb{Z}} H_2(t^n Y) \longrightarrow H_2(\tilde{X})$$

$$\longrightarrow \bigoplus_{n \in \mathbb{Z}} H_1(F) \longrightarrow \bigoplus_{n \in \mathbb{Z}} H_1(t^n Y) \longrightarrow H_1(\tilde{X}) \longrightarrow 0$$

It can be seen that $\bigoplus H_1(t^n Y) = 0$, $\bigoplus H_1(F) = \mathbb{Q}[t^{\pm 1}]^{2g}$ where g is the genus of F, and α is the map

$$\mathbb{Q}[t^{\pm 1}] \longrightarrow (H_2(\Delta) \otimes \mathbb{Q}[t^{\pm 1}]) \oplus \mathbb{Q}[t^{\pm 1}]$$

sending a to $(0, (t-1)a)$. Thus

$$H_2(\tilde{X}) \cong (H_2(\Delta) \otimes \mathbb{Q}[t^{\pm 1}]) \oplus \mathbb{Q} \oplus \mathbb{Q}[t^{\pm 1}]^{2g}.$$

On the other hand, by a Mayer–Vietoris argument for $\tilde W = \tilde X \cup (R \times \mathbb{R})$, we have a long exact sequence

$$H_2(F) \xrightarrow{\beta} H_2(\tilde X) \oplus H_2(R) \longrightarrow H_2(\tilde W)$$
$$\longrightarrow H_1(F) \xrightarrow{\gamma} H_1(\tilde X) \oplus H_1(R) \longrightarrow H_1(\tilde W) \longrightarrow 0$$

where the cokernel of β is exactly

$$H_2(\tilde X) \cong (H_2(\Delta) \otimes \mathbb{Q}[t^{\pm 1}]) \oplus \mathbb{Q}[t^{\pm 1}]^{2g}$$

and γ is the projection $\mathbb{Q}^{2g} \to \mathbb{Q}^g$. From this we obtain $H_1(\tilde W) = 0$ and an exact sequence

$$0 \longrightarrow (H_2(\Delta) \otimes \mathbb{Q}[t^{\pm 1}]) \oplus \mathbb{Q}[t^{\pm 1}]^{2g} \longrightarrow H_2(\tilde W) \longrightarrow \mathbb{Q}^g \longrightarrow 0.$$

Furthermore generators are explicitly identified as follows. Choose 1-cycles $e_1, \ldots, e_{2g}$ in F which form a basis of $H_1(F)$ such that $e_{g+1}, \ldots, e_{2g}$ generate the kernel of $H_1(F) \to H_1(R)$ and $e_1, \ldots, e_g$ are dual to them. We may assume that $i_\pm(e_i \times 0) \subset \partial \Delta$, viewing Δ as a subset of $Y \cong V = \Delta \cup 2$-handle. Appealing to the fact that $\partial \Delta$ is a rational sphere, we can choose 2-chains c_i^+, c_i^- in a collar neighborhood of $\partial \Delta \subset \Delta$ such that $\partial c_i^\pm = i_\pm(e_i \times 0)$ for $i = 1, \ldots, 2g$. From our choice of the e_i, we can assume that there are 2-chains d_i in $R \times \mathbb{R}$ such that $\partial d_i = i_-(e_i \times 0)$ in $\tilde W$ for $i = g+1, \ldots, 2g$. Then $c_i^+ \cdot c_j^- = S(e_i, e_j)$ where S is the Seifert form of F_0. (To verify this, one may appeal to the properties of rational-valued linking number mentioned in [**7**, p. 1169].)

Let
$$v_i = c_i^- \cup -t c_i^+ \quad \text{for } i = 1, \ldots, 2g,$$
$$u_i = c_i^- \cup -d_i \quad \text{for } i = g+1, \ldots, 2g.$$

Then they can be viewed as 2-cycles in $\tilde W$, and from the above computation, it follows that the v_i form a basis of $\mathbb{Q}[t^{\pm 1}]^{2g} \subset H_2(\tilde W)$ and the images of the u_i under $H_2(\tilde W) \to \mathbb{Q}^g$ form a basis of $\mathbb{Q}^g$.

From the intersection data of $c_i^\pm$, the intersection form of $H_2(\tilde W)$ is computed as follows:

$$v_i \cdot v_j = S(e_j, e_i) + S(e_i, e_j) - t^{-1} S(e_j, e_i) - t S(e_i, e_j),$$
$$v_i \cdot u_j = S(e_j, e_i) - t^{-1} S(e_i, e_j).$$

In other words, the restrictions of the intersection form on $\langle v_i \rangle \times \langle v_i \rangle$ and $\langle v_i \rangle \times \langle u_i \rangle$ are represented by

$$(1-t)A + (1 - t^{-1})A^T,$$
$$A^T - t^{-1} A,$$

respectively, where A is the Seifert matrix of F_0 with respect to $\{e_i\}$.

$H_2(W)$ is computed from the above results as follows. Milnor's result on infinite cyclic covers [**37**] gives us a long exact sequence

$$\cdots \longrightarrow H_2(\tilde{W}) \xrightarrow{t-1} H_2(\tilde{W}) \longrightarrow H_2(W) \longrightarrow H_1(\tilde{W}) \xrightarrow{t-1} H_1(\tilde{W}) \longrightarrow \cdots.$$

From this it follows that $H_2(W)$ is isomorphic to the cokernel of $t-1$ on $H_2(\tilde{W})$, since $H_1(\tilde{W}) = 0$. Applying the snake lemma to the commutative diagram

$$\begin{array}{ccccccccc}
0 & \longrightarrow & (H_2(\Delta) \otimes \mathbb{Q}[t^{\pm 1}]) \oplus \mathbb{Q}[t^{\pm 1}]^{2g} & \longrightarrow & H_2(\tilde{W}) & \longrightarrow & \mathbb{Q}^g & \longrightarrow & 0 \\
& & \downarrow{\scriptstyle t-1} & & \downarrow{\scriptstyle t-1} & & \downarrow{\scriptstyle t-1=0} & & \\
0 & \longrightarrow & (H_2(\Delta) \otimes \mathbb{Q}[t^{\pm 1}]) \oplus \mathbb{Q}[t^{\pm 1}]^{2g} & \longrightarrow & H_2(\tilde{W}) & \longrightarrow & \mathbb{Q}^g & \longrightarrow & 0
\end{array}$$

we can see that $H_2(W) \cong H_2(\Delta) \oplus \mathbb{Q}^{2g}$, where the $\mathbb{Q}^{2g}$ factor is generated by the images $\bar{v}_1, \ldots, \bar{v}_g, \bar{u}_{g+1}, \ldots, \bar{u}_{2g} \in H_2(W)$ of $v_1, \ldots, v_g, u_{g+1}, \ldots, u_{2g} \in H_2(\tilde{W})$. Furthermore, the intersection form on $H_2(W)$ is also obtained by plugging in $t = 1$ into the above computation: on $\langle \bar{v}_i \rangle \times \langle \bar{v}_i \rangle$ and $\langle \bar{v}_i \rangle \times \langle \bar{u}_i \rangle$, the intersection forms are given by

$$\left[(1-t)A + (1-t^{-1})A^T \right]_{t=1} = 0,$$
$$\left[A^T - t^{-1}A \right]_{t=1} = A^T - A,$$

respectively. Since the latter is the intersection form on F, it follows that $\{\bar{v}_i\}$ and $\{\bar{u}_i\}$ are dual. Since Δ is a rational (0)-solution, there is a basis $\{x_1, \ldots, x_k, y_1, \ldots, y_k\}$ of $H_2(\Delta)$ such that $\lambda_0(x_i, x_j) = 0$, $\lambda_0(x_i, y_j) = \delta_{ij}$. Then the $\bar{v}_i, x_j, \bar{u}_i, y_j$ form a basis of $H_2(W) \cong H_2(\Delta) \oplus \mathbb{Q}^{2g}$ which satisfies the definition of a rational (0)-solution.

In case of $h = 0.5$, our hypothesis is that Δ is a rational (0.5)-solution of Σ and the Seifert form S of F_0 is metabolic. If

$$H \subset H_1(F) = \operatorname{Coker}\{H_1(\partial F_0) \longrightarrow H_1(F_0)\}$$

is a metabolizer of S, then it can be seen that the pre-image of H under

$$H_1(F; \mathbb{Z}) \longrightarrow H_1(F; \mathbb{Z}) \otimes \mathbb{Q} = H_1(F)$$

is a half-dimensional summand. Choosing a basis of H and dual elements, we obtain a basis $\{e_1, \ldots, e_{2g}\}$ of $H_1(F; \mathbb{Z})$ such that the Seifert matrix A and the intersection matrix $A^T - A$ are of the following form:

$$A = \begin{bmatrix} 0 & * \\ * & * \end{bmatrix}, \quad A^T - A = \begin{bmatrix} 0 & I \\ -I & 0 \end{bmatrix}.$$

We may assume that $\{e_i\}$ is a standard symplectic basis of $H_1(F; \mathbb{Z})$ so that there is a handlebody R bounded by F such that $\langle e_{g+1}, \ldots, e_{2g} \rangle$ is the kernel of $H_1(F) \to H_1(R)$. Now, by performing the above computation using our $\{e_i\}$ and R, W is a rational (0)-solution, and in addition, the intersection form of $H_2(\tilde{W})$ vanishes on the submodule generated by the pre-images $v_1, \ldots, v_g \in H_2(\tilde{W})$ of $\bar{v}_1, \ldots, \bar{v}_g$. Let

$$x'_i = x_i \otimes 1 \in H_2(\Delta) \otimes \mathbb{Q}[t^{\pm 1}] \subset H_2(\tilde{W}).$$

Then $v_1, \ldots, v_g, x'_1, \ldots, x'_k$ are elements in $H_2(\tilde{W})$ which are sent to the basis elements $\bar{v}_1, \ldots, \bar{v}_g, x_1, \ldots, x_k \in H_2(W)$ and the intersection form of $H_2(\tilde{W})$ vanishes on them. It follows that W is a rational (0.5)-solution. This completes the if parts.

The only if part of (1) follows from Proposition 5.7. For the only if part of (2), suppose W is a rational (0.5)-solution of the zero-surgery manifold M. Let c be the complexity of W. Since c is a multiple of the complexity of K, there is a generalized Seifert surface of complexity c, and by attaching parallel copies of the core disks of the added 2-handle in M, we obtain a closed surface F in M. The Thom–Pontryagin construction produces a map $f\colon M \to S^1$ associated to F. It induces $H_1(M;\mathbb{Z}) \to \mathbb{Z}$ sending the meridian of K to c, and thus, it factors through

$$H_1(W;\mathbb{Z}) \longrightarrow H_1(W;\mathbb{Z})/\text{torsion} = \mathbb{Z}.$$

Hence f extends to $W \to S^1$. A transversality argument gives us a properly embedded 3-manifold R in W bounded by F.

Now we modify R so that $H = \operatorname{Ker}\{H_1(F) \to H_1(R)\}$ becomes a metabolizer of the Seifert form of F, by proceeding in a similar way to the proof of [**13**, Proposition 9.2]. We denote the first solvable cover $W^{(1)}$ by $\tilde{W}$ as before. We may assume that the elements $u'_i \in H_2(\tilde{W})$ described in Definition 5.2 are in the image of $H_2(\tilde{W};\mathbb{Z}) \to H_2(\tilde{W})$ by taking multiples of u'_i. Appealing to [**13**, Lemma 7.4], we may assume that the images $u_i \in H_2(W)$ are represented by disjoint surfaces $F_i \subset W$ which are lifted to $\tilde{W}$. Moreover, we may assume that the F_i are disjoint to R by a standard argument removing intersections in the cover $\tilde{W}$; $\tilde{W}$ is obtained by a cut-paste construction using $R \subset W$ so that the intersection of a fixed lift of R and a lift $\tilde{F}_i$ of F_i is a 1-manifold which is null-homologous in $\tilde{F}_i$. We can "surger" R along subsurfaces in $\tilde{F}_i$ to remove the intersection.

Let L be the subgroup in $H_2(W)$ generated by the F_i. By Definition 5.2, $L^\perp = L$ with respect to the intersection form on $H_2(W)$. Given 1-cycles x, y on F representing elements in H, there are 2-chains c, d in R and c' in M such that $\partial c = \partial c' = nx$ and $\partial d = ny$ for some $n \neq 0$. Since $c - c'$ is disjoint to the F_i, $c - c'$ represents an element in $L^\perp$. Since $L = L^\perp$, $m(c - c')$ is a linear combination of the F_i in $H_2(W;\mathbb{Z})$ for some $m \neq 0$. By subtracting this linear combination from mc, we obtain a 2-chain c'' such that $c'' - mc' = 0$. Now the Seifert pairing at (x,y) is given by $n^2 S(x,y) = c' \cdot y^+$ (intersection in M). It is equal to the intersection $c' \cdot d^+$ in W, where d^+ denotes pushoff from R. Since $c'' - mc' = 0$ and the F_i are disjoint from R, $mc' \cdot d^+ = c'' \cdot d^+ = 0$. This proves the claim that H is a metabolizer. $\square$

5.2. Effect of complexity change

In this section we investigate the effect of change of poly-torsion-free-abelian (PTFA) group coefficients on the higher-order Alexander module, Blanchfield form, and von Neumann ρ-invariant. We start by recalling necessary results of Cochran, Orr, and Teichner [13] with a little technical addendum.

5.2.1. Obstructions to admitting a rational solution of a fixed complexity.
In this subsection we discuss an inductive construction of PTFA coefficient systems using the Blanchfield duality, which was used in [13] to define obstructions to admitting a rational solution of a fixed complexity. We will focus on only results that we need to use later. For full details and proofs, see [13].

Let M be a 3-manifold and $\phi \colon \pi_1(M) \to \Gamma$ be a homomorphism into a PTFA group Γ. Let $\mathcal{K} = \mathbb{Q}\Gamma(\mathbb{Q}\Gamma - \{0\})^{-1}$ be the skew field of quotients of the Ore domain $\mathbb{Q}\Gamma$ which is obtained by inverting nonzero elements from right. Let $\mathcal{R}$ be a subring of $\mathcal{K}$ containing $\mathbb{Q}\Gamma$. Then, the homology group $H_*(M; \mathcal{R})$ with $\mathcal{R}$-coefficient is defined. We recall its definition for later use. Let X be the regular cover of M associated to $\pi_1(M) \to \Gamma$. The cellular chain complex $C_*(X; \mathbb{Z})$ becomes a $\mathbb{Z}[\Gamma]$-module via the covering transformation action of Γ. $H_*(M; \mathcal{R})$ is defined to be the homology of the chain complex

$$C_*(M; \mathcal{R}) = C_*(X; \mathbb{Z}) \otimes_{\mathbb{Z}[\Gamma]} \mathcal{R}.$$

The associated *(rational) Alexander module* is defined to be the homology module $\mathcal{A} = H_1(M; \mathcal{R})$. There is a nondegenerated linking form

$$B\ell \colon \mathcal{A} \times \mathcal{A} \longrightarrow \mathcal{K}/\mathcal{R}$$

which is called the *Blanchfield form* [13]. For later use we give a geometric description of $B\ell$. Given 1-cycles x and y in $C_1(M; \mathcal{R})$, there is a 2-cycle u in $C_2(M; \mathcal{R})$ such that $\partial u = ax$ for some nonzero $a \in \mathcal{R}$ since $\mathcal{A}$ is a torsion $\mathcal{R}$-module (e.g., see [13]). Then

$$B\ell(x, y) = \frac{1}{a} \cdot I(u, y) + \mathcal{R} \in \mathcal{K}/\mathcal{R}$$

where $I(u, y)$ denotes the $\mathcal{R}$-valued twisted intersection number of u and y.

By a universal coefficient spectral sequence argument and a standard interpretation of the first group cohomology as the set of derivations, we have

$$\operatorname{Hom}(\mathcal{A}, \mathcal{K}/\mathcal{R}) \cong H^1(M; \mathcal{K}/\mathcal{R}) \cong \frac{\{\text{derivations } \pi_1(M) \to \mathcal{K}/\mathcal{R}\}}{\{\text{principal derivations}\}}.$$

Given an element $x \in \mathcal{A}$, the adjoint map $\mathcal{A} \to \mathcal{K}/\mathcal{R}$ sending y to $B\ell(y, x)$ gives rise to a derivation $d \colon \pi_1(M) \to \mathcal{K}/\mathcal{R}$ which is unique up to principal derivations. By the universal property of the semidirect product, it induces a homomorphism

$$\varphi = \varphi(x, \phi) \colon \pi_1(M) \longrightarrow \mathcal{K}/\mathcal{R} \rtimes \Gamma$$

given by $\varphi(g) = (d(g), \phi(g))$. In [**13**] it was shown that φ is well-defined up to $\mathcal{K}/\mathcal{R}$-conjugation.

This construction is applied inductively to construct coefficient systems of the surgery manifold of a knot over the following PTFA groups.

DEFINITION 5.12. The *n-th rationally universal group* Γ_n is defined inductively by
$$\Gamma_0 = \mathbb{Z}, \quad \Gamma_{n+1} = \mathcal{K}_n/\mathcal{R}_n \rtimes \Gamma_n \quad (n \geq 0)$$
where $\mathcal{K}_n$ is the skew field of quotients of $\mathbb{Q}\Gamma_n$ and
$$\mathcal{R}_n = \mathbb{Q}[\Gamma_n](\mathbb{Q}[\Gamma_n, \Gamma_n] - 0)^{-1} \subset \mathcal{K}_n.$$
(In [**13**] our Γ_n was denoted by Γ_n^U.)

Henceforth we view Γ_0 as the multiplicative infinite cyclic group $\langle t \rangle$ generated by t.

Suppose that K is a knot in a rational sphere with vanishing $\mathbb{Q}/\mathbb{Z}$-valued self-linking. Let M be the result of surgery along the zero-framing of K on the ambient space. Fix a positive multiple c of the complexity of K. We construct Γ_n-coefficient systems ϕ_n on M, which depend on the choice of c. Let
$$\phi_0 \colon \pi_1(M) \longrightarrow \Gamma_0 = \langle t \rangle$$
be the homomorphism sending the (positively oriented) meridian of K to $t^c \in \langle t \rangle$, which (uniquely) exists by our choice of c. Suppose $\phi_n \colon \pi_1(M) \to \Gamma_n$ has been defined. Choosing $x_n \in \mathcal{A}_n = H_1(M; \mathcal{R}_n)$, a new coefficient system
$$\phi_{n+1} = \phi_{n+1}(x_n, \phi_n) \colon \pi_1(M) \longrightarrow \Gamma_{n+1}$$
is induced as discussed above.

Given a closed 3-manifold M and a group homomorphism $\phi \colon \pi_1(M) \to G$, there defined the *von Neumann signature invariant* $\rho(M, \phi) \in \mathbb{R}$ (see Cheeger–Gromov [**9**]). The following theorem of Cochran–Orr–Teichner [**13**] states that for a certain choice of x_n, $\rho(M, \phi_n)$ gives an obstruction to being rationally $(n.5)$-solvable via a rational $(n.5)$-solution of *complexity c*.

THEOREM 5.13 (c.f., Theorem 4.6 of [**13**]). *Suppose K is a rational knot with vanishing $\mathbb{Q}/\mathbb{Z}$-valued self-linking, M is the surgery manifold of K, and $\phi_0 \colon \pi_1(M) \to \Gamma_0$ is the homomorphism sending the meridian of K to t^c, where c is a positive multiple of the complexity of K. If W is a rational (n)-solution of complexity c for K, then the following statements hold:*

(0) *$\phi_0 \colon \pi_1(M) \to \Gamma_0$ factors through $\pi_1(W)$. It gives rise to the Alexander module $\mathcal{A}_0 = H_1(M; \mathcal{R}_0)$ and the Blanchfield form*
$$B\ell_0 \colon \mathcal{A}_0 \times \mathcal{A}_0 \longrightarrow \mathcal{K}_0/\mathcal{R}_0.$$

(0.5) *$\rho(M, \phi_0) = 0$.*

(1) *$P_0 = \mathrm{Ker}\{\mathcal{A}_0 \to H_1(W; \mathcal{R}_0)\}$ is self-annihilating with respect to $B\ell_0$, and for any $x_0 \in P_0$, the induced coefficient system*
$$\phi_1 = \phi_1(x_0, \phi_0) \colon \pi_1(M) \longrightarrow \Gamma_1$$

factors through $\pi_1(W)$. It gives rise to the Alexander module $\mathcal{A}_1 = H_1(M;\mathcal{R}_1)$ and the Blanchfield form

$$B\ell_1 \colon \mathcal{A}_1 \times \mathcal{A}_1 \longrightarrow \mathcal{K}_1/\mathcal{R}_1.$$

(1.5) $\rho(M, \phi_1) = 0$.

$\vdots$

(n) $P_{n-1} = \mathrm{Ker}\{\mathcal{A}_{n-1} \to H_1(W;\mathcal{R}_{n-1})\}$ *is self-annihilating with respect to* $B\ell_{n-1}$, *and for any* $x_{n-1} \in P_{n-1}$, *the induced coefficient system*

$$\phi_n = \phi_n(x_{n-1}, \phi_{n-1}) \colon \pi_1(M) \longrightarrow \Gamma_n$$

factors through $\pi_1(W)$. It gives rise to the Alexander module $\mathcal{A}_n = H_1(M;\mathcal{R}_n)$ and the Blanchfield form

$$B\ell_n \colon \mathcal{A}_n \times \mathcal{A}_n \longrightarrow \mathcal{K}_n/\mathcal{R}_n.$$

In addition, if W is a rational $(n.5)$-solution of complexity c, then the following statement holds:

(n.5) $\rho(M, \phi_n) = 0$.

REMARK 5.14. In the original work of Cochran, Orr, and Teichner [**13**], they discussed this result under the following restrictions:

(1) They considered rational solvability of knots in S^3 only. We generalizes it for rational knots admitting well-defined zero-surgery manifolds.
(2) They stated the hypothesis on ϕ_0 in terms of the notion of "multiplicity", rather than complexity. In particular, they considered only the case that the extension $\pi_1(W) \to \mathbb{Z}$ of $\phi_0 \colon \pi_1(M) \to \mathbb{Z}$ is surjective. We do not require it.

In spite of this, their original proof works for Theorem 5.13 without any substantial modification. We do not repeat the details.

We emphasize again that the obstructions in Theorem 5.13 depends on the choice of c. To handle all the possible values of c, we investigate the effect of change of c in the next subsection.

5.2.2. Change of coefficient systems. In this subsection we observe naturality of higher order Alexander modules, Blanchfield forms, and associated ρ-invariants with respect to coefficients.

Suppose that $\phi \colon \pi_1(M) \to \Gamma$ and $\phi' \colon \pi_1(M) \to \Gamma'$ are PTFA coefficient systems and $h \colon \Gamma \to \Gamma'$ is an injection making the following diagram commute:

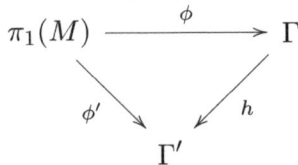

Note that in this case the von Neumann invariants $\rho(M,\phi)$ and $\rho(M,\phi')$ are the same by the following result.

PROPOSITION 5.15 (Subgroup property). *If $\phi\colon \pi_1(M) \to G$ is a homomorphism and $i\colon G \to G'$ is an injection, then $\rho(M,\phi) = \rho(M, i \circ \phi)$.*

This is due to Cheeger–Gromov [**9**]. See also [**13**, Proposition 5.13].

We investigate the relationship of induced coefficient systems obtained from ϕ and ϕ'. Let $\mathcal{K}$ be the skew field of quotients of $\mathbb{Q}\Gamma$ and $\mathcal{R}$ be a subring such that $\mathbb{Q}\Gamma \subset \mathcal{R} \subset \mathcal{K}$ as before, and $\mathbb{Q}\Gamma' \subset \mathcal{R}' \subset \mathcal{K}'$ similarly. Let $\mathcal{A} = H_1(M;\mathcal{R})$ and $\mathcal{A}' = H_1(M;\mathcal{R}')$ be the Alexander modules and $B\ell$ and $B\ell'$ be the Blanchfield forms associated to ϕ and ϕ', respectively. We assume that the induced homomorphism $\mathcal{K} \to \mathcal{K}'$ sends $\mathcal{R}$ into $\mathcal{R}'$ so that $\mathcal{R}'$ can be viewed as an $\mathcal{R}$-module.

THEOREM 5.16. *If $\mathcal{R}$ is a PID, then the followings hold:*

(1) $\mathcal{A}' = \mathcal{A} \otimes_{\mathcal{R}} \mathcal{R}'$.
(2) *The Blanchfield form $B\ell'\colon \mathcal{A}' \times \mathcal{A}' \to \mathcal{K}'/\mathcal{R}'$ is given by*

$$B\ell'(x \otimes a, y \otimes b) = a \cdot B\ell(x,y)^h \cdot \bar{b}$$

where $B\ell(x,y)^h$ is the image of $B\ell(x,y)$ under the induced homomorphism $\mathcal{K}/\mathcal{R} \to \mathcal{K}'/\mathcal{R}'$.
(3) *For $x' = x \otimes 1 \in \mathcal{A} \otimes \mathcal{R}' = \mathcal{A}'$, the coefficient system*

$$\varphi' = \varphi'(x', \phi')\colon \pi_1(M) \longrightarrow \mathcal{K}'/\mathcal{R}' \rtimes \Gamma'$$

induced by x' and ϕ' is given by $\varphi' = \bar{h} \circ \varphi$, where $\bar{h}$ is the homomorphism induced by h.

$$\begin{array}{ccc}
\pi_1(M) & \xrightarrow{\varphi} & \mathcal{K}/\mathcal{R} \rtimes \Gamma \\
& \searrow{\varphi'} \quad \swarrow{\bar{h}} & \\
& \mathcal{K}'/\mathcal{R}' \rtimes \Gamma' &
\end{array}$$

(4) *$\bar{h}$ is injective if and only if the pre-image of $\mathcal{R}'$ under $\mathcal{K} \to \mathcal{K}'$ is exactly $\mathcal{R}$. In this case, we have*

$$\rho(M,\varphi) = \rho(M,\varphi').$$

PROOF. Denote the regular coverings of M associated to ϕ and ϕ' by X and X', respectively. Since X' is the disjoint union of copies of X indexed by cosets of $h(\Gamma)$ in Γ', the cellular chain complex of X' is given by

$$C_*(X';\mathbb{Z}) = C_*(X;\mathbb{Z}) \otimes_{\mathbb{Z}\Gamma} \mathbb{Z}\Gamma'.$$

Thus the $\mathcal{R}'$-coefficient chain complex $C_*(M;\mathcal{R}')$ can be computed in terms of $C_*(M;\mathcal{R})$:

$$\begin{aligned} C_*(M;\mathcal{R}') &= C_*(X';\mathbb{Z}) \otimes_{\mathbb{Z}\Gamma'} \mathcal{R}' \\ &= C_*(X;\mathbb{Z}) \otimes_{\mathbb{Z}\Gamma} \mathcal{R}' \\ &= (C_*(X;\mathbb{Z}) \otimes_{\mathbb{Z}\Gamma} \mathcal{R}) \otimes_{\mathcal{R}} \mathcal{R}' \\ &= C_*(M;\mathcal{R}) \otimes_{\mathcal{R}} \mathcal{R}'. \end{aligned}$$

Since $\mathcal{R}' \subset \mathcal{K}'$ and $\mathcal{R}$ injects into $\mathcal{K}'$, $\mathcal{R}'$ is $\mathcal{R}$-torsion free. Since $\mathcal{R}$ is a PID, $\mathcal{R}'$ is $\mathcal{R}$-free (one may appeal to a noncommutative version of the structure theorem of modules over a PID, e.g., see [15]). Hence (1) follows from the universal coefficient theorem.

Furthermore the $\mathcal{R}'$-valued intersection form

$$(C_1(M;\mathcal{R}) \otimes \mathcal{R}') \times (C_2(M;\mathcal{R}) \otimes \mathcal{R}') \longrightarrow \mathcal{R}'$$

is given by $(x \otimes a, y \otimes b) \to \bar{a} \cdot (x \cdot y)^h \cdot b$ where $(x \cdot y)^h$ is the image of the $\mathcal{R}$-valued intersection of x and y under $\mathcal{R} \to \mathcal{R}'$. Thus (2) follows from the geometric description of the Blanchfield form discussed in the previous subsection.

From (2), the derivation $\pi_1(M) \to \mathcal{K}'/\mathcal{R}'$ associated to $x \otimes 1 \in \mathcal{A}'$ is the composition

$$\pi_1(M) \xrightarrow{d} \mathcal{K}/\mathcal{R} \longrightarrow \mathcal{K}'/\mathcal{R}'$$

where d is the derivation associated to $x \in \mathcal{A}$. Thus by the definition, $\varphi' = \bar{h} \circ \varphi$. This shows (3).

For (4), observe that

$$\bar{h} \colon \mathcal{K}/\mathcal{R} \rtimes \Gamma \longrightarrow \mathcal{K}'/\mathcal{R}' \rtimes \Gamma'$$

is an injection if and only if so is $\mathcal{K}/\mathcal{R} \to \mathcal{K}'/\mathcal{R}'$, since $h \colon \Gamma \to \Gamma'$ is injective by the hypothesis. The last conclusion follows from the subgroup property. □

Applying Theorem 5.16 inductively to the construction of rationally universal coefficient systems discussed in the previous subsection, we obtain the following corollary: suppose c is a multiple of the complexity of K, and

$$\phi_0, \phi_0' \colon \pi_1(M) \longrightarrow \Gamma_0 = \langle t \rangle$$

are homomorphisms sending the meridian of K to t^c and t^{rc}, respectively. Let $h_n \colon \Gamma_n \to \Gamma_n$ be the homomorphism induced by $h_0 \colon \Gamma_0 \to \Gamma_0$ sending t to t^r. h_n gives rise to another $\mathcal{R}_n$-bimodule structure on $\mathcal{R}_n$ via $r \cdot x \cdot s = h_n(r)xs$. For a right $\mathcal{R}_n$-module $\mathcal{M}$, we denote the tensor product of $\mathcal{M}$ and $\mathcal{R}_n$ with this module structure by $\mathcal{M} \otimes_{h_n} \mathcal{R}_n$.

COROLLARY 5.17.
(0) *Suppose $\mathcal{A}_0$ is the Alexander module associated to ϕ_0 and $B\ell_0$ is the Blanchfield form on $\mathcal{A}_0$. Then*
 – *The coefficient system ϕ_0' is given by $\phi_0' = h_0 \circ \phi_0$.*

- The Alexander module $\mathcal{A}'_0$ associated to ϕ'_0 is given by
$$\mathcal{A}'_0 = \mathcal{A}_0 \otimes_{h_0} \mathcal{R}_0.$$
- The Blanchfield form $B\ell'_0$ on $\mathcal{A}'_0$ is given by
$$B\ell'_0(x \otimes a, y \otimes b) = a \cdot B\ell_0(x,y)^{h_0} \cdot \bar{b}.$$
- $\rho(M, \phi_0) = \rho(M, \phi'_0)$.

(1) Suppose $\phi_1 = \phi_1(x_0, \phi_0) \colon \pi_1(M) \to \Gamma_1$ is the coefficient system corresponding to $x_0 \in \mathcal{A}_0$, $\mathcal{A}_1$ is the associated Alexander module, and $B\ell_1$ is the Blanchfield form on $\mathcal{A}_1$. Then for $x'_0 = x_0 \otimes 1 \in \mathcal{A}'_0$,
  - The coefficient system $\phi'_1 = \phi'_1(x'_0, \phi'_0)$ is given by $\phi'_1 = h_1 \circ \phi_1$.
  - The Alexander module $\mathcal{A}'_1$ associated to ϕ'_1 is given by
$$\mathcal{A}'_1 = \mathcal{A}_1 \otimes_{h_1} \mathcal{R}_1.$$
  - The Blanchfield form $B\ell'_1$ on $\mathcal{A}'_1$ is given by
$$B\ell'_1(x \otimes a, y \otimes b) = a \cdot B\ell_1(x,y)^{h_1} \cdot \bar{b}.$$
  - $\rho(M, \phi_1) = \rho(M, \phi'_1)$.

$\vdots$

(n) Suppose $\phi_n = \phi_n(x_{n-1}, \phi_{n-1}) \colon \pi_1(M) \to \Gamma_n$ is the coefficient system corresponding to $x_{n-1} \in \mathcal{A}_{n-1}$, $\mathcal{A}_n$ is the associated Alexander module, and $B\ell_n$ is the Blanchfield form on $\mathcal{A}_n$. Then for $x'_{n-1} = x_{n-1} \otimes 1 \in \mathcal{A}'_{n-1}$,
  - The coefficient system $\phi'_n = \phi'_n(x'_{n-1}, \phi'_{n-1})$ is given by $\phi'_n = h_n \circ \phi_n$.
  - The Alexander module $\mathcal{A}'_n$ associated to ϕ'_n is given by
$$\mathcal{A}'_n = \mathcal{A}_n \otimes_{h_n} \mathcal{R}_n.$$
  - The Blanchfield form $B\ell'_n$ on $\mathcal{A}'_n$ is given by
$$B\ell'_n(x \otimes a, y \otimes b) = a \cdot B\ell_n(x,y)^{h_n} \cdot \bar{b}.$$
  - $\rho(M, \phi_n) = \rho(M, \phi'_n)$.

$\vdots$

PROOF. The only one thing we need to check is whether the homomorphism $f \colon \mathcal{K}_n \to \mathcal{K}_n$ induced by h_n satisfies $f^{-1}(\mathcal{R}_n) = \mathcal{R}_n$. This implies that h_n is injective for all n. For this we appeal to the fact that
$$\mathcal{R}_n = \mathbb{Q}\Gamma_n(\mathbb{Q}[\Gamma_n, \Gamma_n] - \{0\})^{-1}$$
is isomorphic the Laurent polynomial ring $\mathbb{K}[t^{\pm 1}]$ over the skew field of quotients $\mathbb{K}$ of $\mathbb{Q}[\Gamma_n, \Gamma_n]$, where t is represented by a generator of $\Gamma_0 = \mathbb{Z}$ (see [**13**]). Thus $\mathcal{K}_n$ is isomorphic to the skew field of rational functions $\mathbb{K}(t)$, and the concerned homomorphism $f \colon \mathbb{K}(t) \to \mathbb{K}(t)$ is given by
$$\sum t^i \cdot a_i \longrightarrow \sum t^{ri} \cdot (a_i)^{h_n}$$

where $a \to a^{h_n}$ denotes the homomorphism on $\mathbb{K}$ induced by h_n. Combining this with the long-division algorithm, it follows that, for any $P(t), Q(t) \in \mathbb{K}[t^{\pm 1}]$, if $f(P(t))$ divides $f(Q(t))$ then $P(t)$ divides $Q(t)$. This shows

$$f^{-1}(\mathbb{K}[t^{\pm 1}]) = \mathbb{K}[t^{\pm 1}].$$ □

From Corollary 5.17, we can see that if the metabolizer $P_n \subset \mathcal{A}_n$ in Theorem 5.13 can be controlled in an appropriate way as c varies, then we can choose elements $x_n \in P_n$ such that the value of the associated ρ-invariant in Theorem 5.13 is independent of c. In this case a single ρ-invariant would obstruct the existence of rational solutions of *any* complexity. The next section is devoted to a construction of examples for which we can control the first metabolizer P_0 as desired.

5.3. Realization of Alexander modules by ribbon knots

In this section we discuss realization of certain classical Alexander modules and Blanchfield forms by ribbon knots. Recall that the classical Alexander module of a knot K in S^3 is defined to be $H_1(S^3 - K; \mathbb{Z}[t^{\pm 1}])$ where the $\mathbb{Z}[t^{\pm 1}]$-coefficient system is induced by $\pi_1(S^3 - K) \to H_1(S^3 - K) = \langle t \rangle$ sending the meridian to t.

THEOREM 5.18. *For any polynomial $P(t)$ with integer coefficients such that $P(1) = \pm 1$ and $P(t^{-1}) = P(t)$ up to multiplication by $\pm t^n$, there is a ribbon knot K in S^3 whose classical Alexander module is $\mathbb{Z}[t^{\pm 1}]/\langle P(t)^2 \rangle$.*

We remark that some special cases of Theorem 5.18 were considered by Kim [**28**] and Friedl [**18**]. Theorem 5.18 generalizes their ad-hoc methods. We also remark that there are well-known realization results of Alexander polynomials by slice knots. Since the author could not find the necessary realization result of Alexander modules in the literature, he gives a proof of Theorem 5.18 for concreteness.

PROOF. For notational convenience we may assume that $P(t)$ is of the form

$$P(t) = a_g t^g + \cdots + a_1 t + a_0 + a_1 t^{-1} + \cdots + a_g t^{-g}$$

and $P(1) = 1$ without any loss of generality.

The knot K is obtained by a surgery construction which is similar to [**31**, **16**]. See Figure 1. We perform $(+1)$, (-1), and (-1)-surgery on S^3 along the curves α_1, α_2, and α_3 illustrated in Figure 1, respectively. In Figure 1 $(\pm a_i)$ denotes the number of full twists. α_1 may be viewed as a knot obtained by g band sum operations on a link with $(g+1)$ unknotted components $C_0, \ldots, C_g$ satisfying $lk(C_0, C_i) = a_i$ and $lk(C_i, C_j) = 0$ for $i, j > 0$. The band joining C_0 and C_i wraps i times the unknotted circle K_0 for $i = 1, \ldots, g$. Similarly for α_2; observe the symmetry of Figure 1 with respect to the reflection about a horizontal mirror. The result of surgery along α_1, α_2, and α_3 is again S^3, and K_0 becomes a (nontrivial) knot K in S^3.

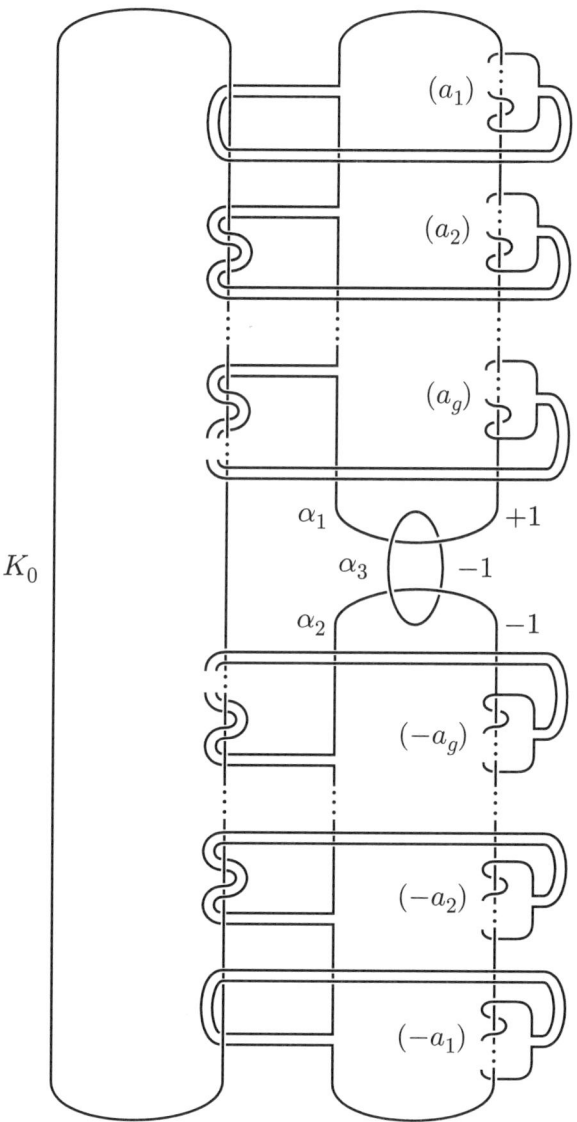

FIGURE 1

We will show that this knot K has the desired properties. First we compute the classical Alexander module of K. The infinite cyclic cover of the complement of the unknotted circle K_0 is $\mathbb{R}^3$, and the pre-image of each α_i consists of infinitely many simple closed curves $t^j \beta_i$ where t denotes the covering transformation corresponding to the meridian of K_0. Figure 2 illustrates these curves in $\mathbb{R}^3$.

The infinite cyclic cover of the complement of K is obtained by performing surgery along the curves $t^j \beta_i$, $j \in \mathbb{Z}$, $i = 1, 2, 3$. The surgery framings in

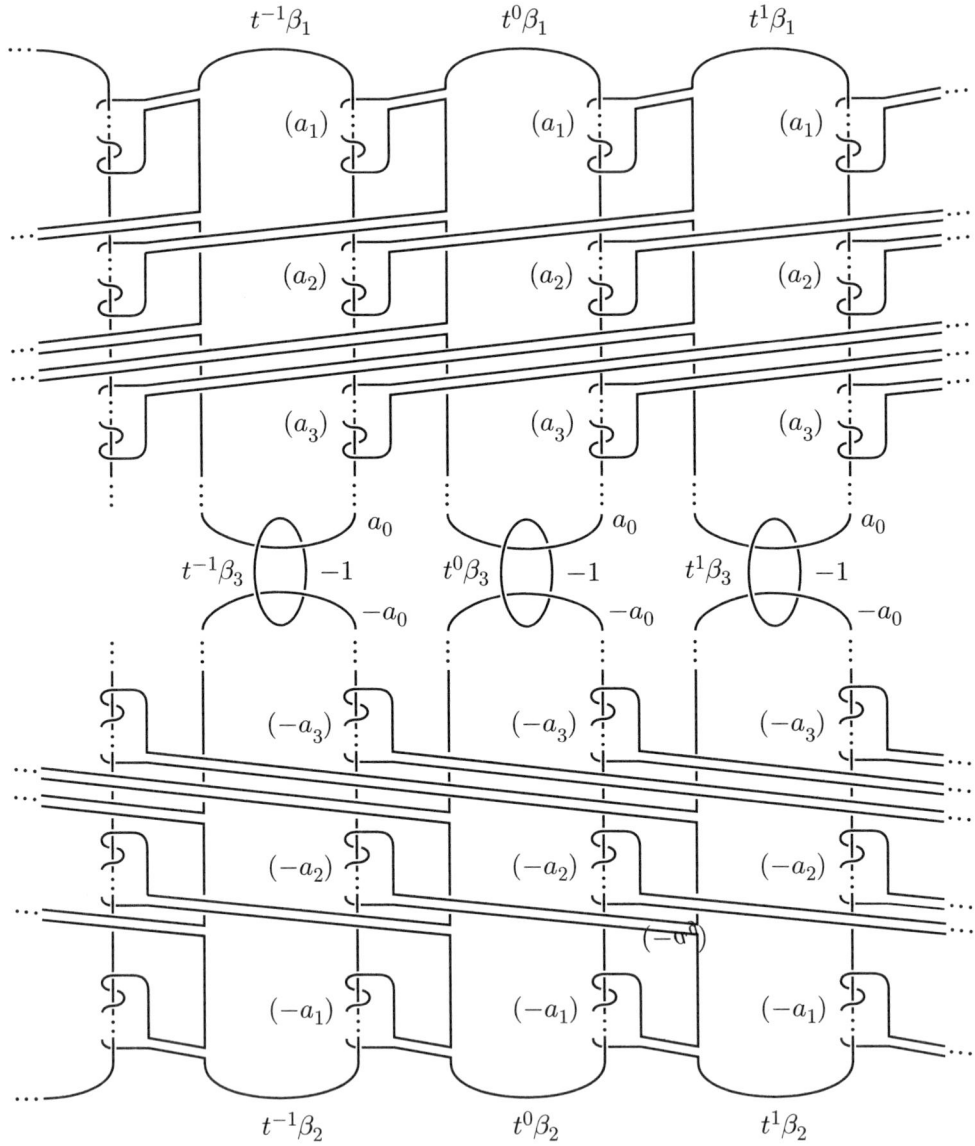

FIGURE 2

Figure 2 can be verified using the equation
$$(\text{framing on } \beta_i) + \sum_{j \neq 0} \text{lk}(\beta_i, t^j \beta_i) = (\text{framing on } \alpha_i).$$
Here we need the fact that $P(1) = 2(a_g + \cdots + a_1) + a_0 = 1$ and $\text{lk}(\beta_i, t^j \beta_i) = a_{|j|}$ if $i = 1$, $-a_{|j|}$ if $i = 2$.

By the Alexander duality, the exterior X of $\bigcup_{i,j} t^j \beta_i$ has $H_1(X; \mathbb{Z}[t^{\pm 1}]) = \mathbb{Z}[t^{\pm 1}]^3$ which is generated by the meridian e_i of β_i, $i = 1, 2, 3$. Filling X

with copies of $S^1 \times D^2$, we obtain the infinite cyclic cover of the complement of K, and its first $\mathbb{Z}[t^{\pm 1}]$-homology is the quotient of $H_1(X; \mathbb{Z}[t^{\pm 1}])$ by the $\mathbb{Z}[t^{\pm 1}]$-submodule generated by the parallels of the β_i corresponding to the framings, $i = 1, 2, 3$. From mutual linkings and framings of $t^j \beta_i$, it follows that the relations from surgery gives us a presentation matrix

$$\begin{bmatrix} P(t) & 0 & 1 \\ 0 & -P(t) & 1 \\ 1 & 1 & -1 \end{bmatrix}$$

of the classical Alexander module of K.

Adding the last row to the first and second rows, we can eliminate the last row and column from the presentation. This gives us a new matrix

$$\begin{bmatrix} P(t)+1 & 1 \\ 1 & -P(t)+1 \end{bmatrix}.$$

Adding $(P(t) - 1)$ times the first row to the second row, we can eliminate the first row and the second column, and the resulting 1×1 matrix gives us the module $\mathbb{Z}[t^{\pm 1}]/\langle P(t)^2 \rangle$ as desired.

Now it remains to show that K is ribbon. For this purpose we transform Figure 1 as in Figure 3. First, since the simple closed curves α_1 and α_2 in Figure 1 are unknotted, Figure 1 can be isotoped to the first Kirby diagram in Figure 3, where T represents a tangle and $-T$ is its mirror image with respect to a vertical mirror. Next, by performing "Rolfsen twist" on α_1 and α_2, we obtain the second Kirby diagram in Figure 3. By this, as illustrated in the second diagram, the curve α_3 becomes the boundary of a 2-disk D^2, and K_0 becomes a knotted circle K_1 in S^3. Observe that K_1 bounds a ribbon disk D in S^3 since it is a connected sum of a knot and its mirror image. Furthermore, the intersection of D and the 2-disk D^2 consists of disjoint arcs in the interior of D^2. Finally we perform another Rolfsen twist along ∂D^2. It gives us our knot K illustrated in the last diagram in Figure 3. In fact, K is obtained from K_1 by cutting S^3 along D^2 and pasting along a full rotation on D^2. This operation transforms the ribbon disk D of K_1 into a ribbon disk of K. It completes the proof. $\square$

REMARK 5.19. By a straightforward computation, it can also be seen that the Blanchfield form of K

$$B\ell \colon \mathcal{A} \times \mathcal{A} \longrightarrow \mathbb{Q}(t)/\mathbb{Z}[t^{\pm 1}]$$

is given by

$$B\ell(f(t), g(t)) = \frac{f(t)g(t)(P(t)-1)}{P(t)^2} \in \mathbb{Q}(t)/\mathbb{Z}[t^{\pm 1}]$$

where the Alexander module $\mathcal{A}$ is identified with $\mathbb{Z}[t^{\pm 1}]/\langle P(t)^2 \rangle$ in such a way that the meridian e_1 of α_1 is identified with 1, as in the above proof.

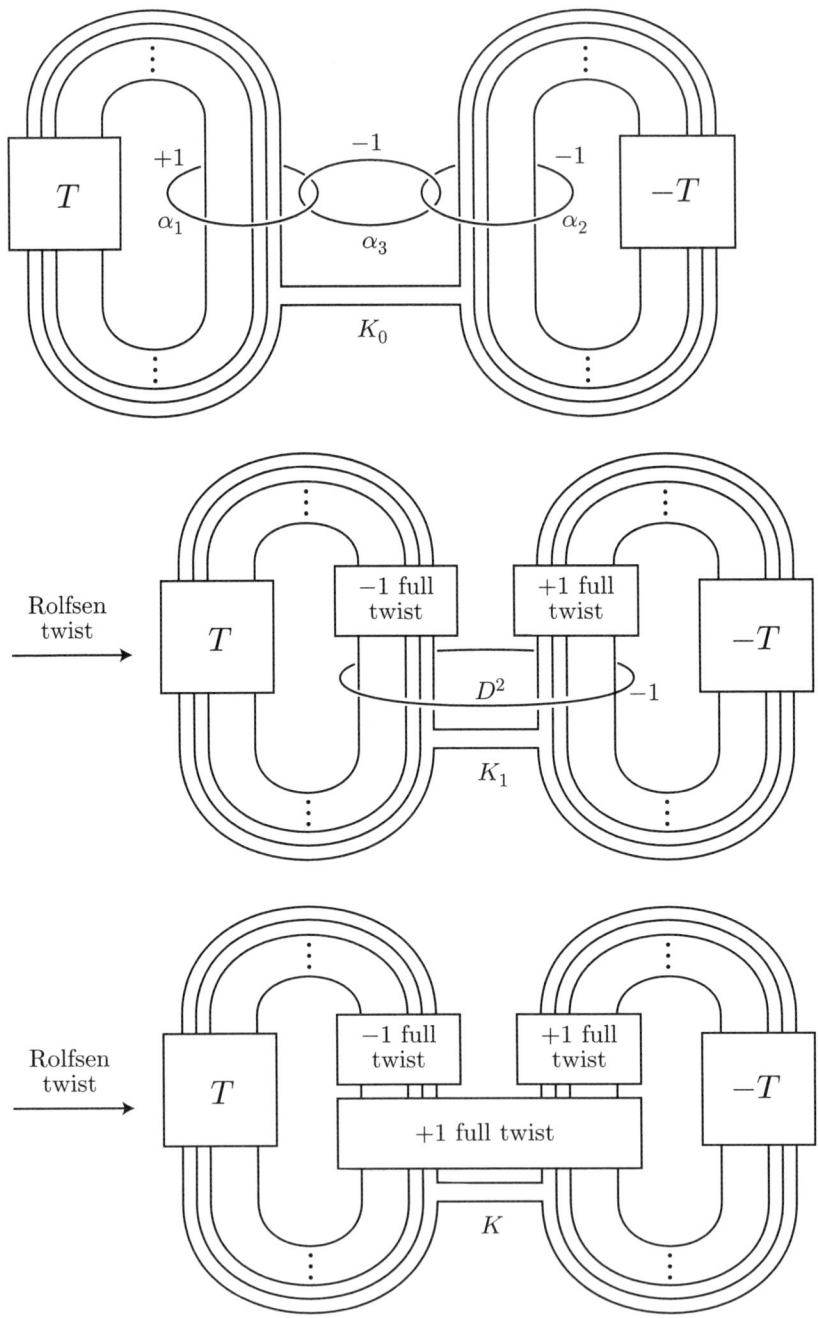

FIGURE 3

5.4. Knots which are not rationally (1.5)-solvable

In this section we construct a family of knots in S^3 which have metabolic Seifert matrices but are not rational slice knots. In fact we will show that

5.4. KNOTS WHICH ARE NOT RATIONALLY (1.5)-SOLVABLE

for each knot in the family a single von Neumann ρ-invariant obstructs the existence of rational solutions of *any complexity*.

In the construction we need the following special case of Proposition 3.18, which will play a crucial role in controlling the configuration of metabolizer in the rational Alexander module for an arbitrary complexity.

LEMMA 5.20 (A special case of Proposition 3.18). *Suppose* $\lambda(t) = at^2 - (2a+1)t + a$ *where a is an odd prime. Then $\lambda(t^r)$ is irreducible for any positive integer r.*

In the following statement, M is the zero-surgery manifold of a knot K in S^3 and $\mathcal{A}_0$ is its rational Alexander module associated to $\phi_0 \colon \pi_1(M) \to \Gamma_0 = \langle t \rangle$ sending the meridian to t.

PROPOSITION 5.21. *Suppose $\lambda(t)$ is a polynomial satisfying Lemma 5.20 and K is a knot in S^3 such that $\mathcal{A}_0 = \mathbb{Q}[t^{\pm 1}]/\langle \lambda(t)^2 \rangle$. If K has a rational (1.5)-solution of any complexity, then for any element x_0 of the form*

$$x_0 = f(t)\lambda(t) + \langle \lambda(t)^2 \rangle \in \mathcal{A}_0$$

where $f(t)$ is a polynomial, the von Neumann invariant $\rho(M, \phi_1)$ associated to

$$\phi_1 = \phi_1(x_0, \phi_0) \colon \pi_1(M) \longrightarrow \Gamma_1$$

vanishes.

PROOF. Suppose that there is a rational (1.5)-solution. Let denote its complexity by c. Let $\phi_0' \colon \pi_1(M) \to \langle t \rangle$ be the homomorphism sending the meridian of K to t^c and $h_0 \colon \Gamma_0 \to \Gamma_0$ be the map $t \to t^c$ as in Section 5.2. By Corollary 5.17, the Alexander module $\mathcal{A}_0'$ associated to ϕ_0' is given by

$$\mathcal{A}_0' \cong \mathcal{A}_0 \otimes_{h_0} \mathcal{R}_0 \cong \mathbb{Q}[t^{\pm 1}]/\langle \lambda(t^c)^2 \rangle.$$

Since $\lambda(t^c)$ is irreducible, $\mathcal{A}_0'$ has a unique proper submodule $\lambda(t^c)\mathcal{A}_0' = P \otimes_{h_0} \mathcal{R}_0$ where $P = \lambda(t)\mathcal{A}_0 \subset \mathcal{A}_0$.

By Theorem 5.13, there is a self-annihilating submodule $P' \in \mathcal{A}_0'$ such that $\rho(M, \phi_1') = 0$ for any $\phi_1' \colon \pi_1(M) \to \Gamma_1$ associated to an element in P'. Since P' is a proper submodule, P' must be equal to $P \otimes_{h_0} \mathcal{R}_0$. Since $x_0 = f(t)\lambda(t) + \langle \lambda(t)^2 \rangle$ is contained in P, $x_0' = x_0 \otimes 1$ is contained in P'. Let

$$\phi_1' = \phi_1'(x_0', \phi_0') \colon \pi_1(M) \longrightarrow \Gamma_1.$$

Then from Corollary 5.17, it follows that $\rho(M, \phi_1) = \rho(M, \phi_1') = 0$. □

For any polynomial $\lambda(t)$ satisfying Lemma 5.20, we can choose a ribbon knot with classical Alexander module $\mathbb{Z}[t^{\pm 1}]/\langle \lambda(t)^2 \rangle$ by appealing to Theorem 5.18. This knot satisfies the hypothesis of Proposition 5.21.

We will modify this knot, without altering the Alexander module, to realize a nontrivial value of the concerned ρ-invariant. For this purpose, as in [**13, 14, 28**], we use a well-known construction which is sometimes called "satellite construction", "grafting construction", or "genetic modification". Let K' be a knot in S^3 with zero-surgery manifold M', and η be an unknotted simple closed curves in $S^3 - K'$. Let J be another knot in S^3. We modify K' by "tying J along η" as follows. Remove an open tubular neighborhood U_η of η, and fill it in with the exterior E_J of J along an orientation reversing homeomorphism between their boundaries which identifies the meridian and the zero-linking longitude of J with the zero-linking longitude and the meridian of η, respectively. Then we obtain again S^3, but K' becomes a new knot, say K. The zero-surgery manifold M of K is given by $M = (M' - U_\eta) \cup_\partial E_J$. It is known that the ρ-invariant of K is expressed in terms of the ρ-invariant of K' and the signature function of J. Let

$$\rho(J) = \int \sigma_w(J)\, dw$$

be the integral of the knot signature function

$$\sigma_w(J) = \text{sign}\left((1-w)A + (1-w^{-1})A^T\right)$$

of J over the complex unit circle normalized to length one, where A is a Seifert matrix of J.

LEMMA 5.22 ([**14, 28**]). *If ϕ' is a homomorphism of $\pi_1(M')$ into a PTFA group Γ, then there is a unique homomorphism $\phi\colon \pi_1(M) \to \Gamma$ such that the compositions*

$$\pi_1(M' - U_\eta) \longrightarrow \pi_1(M) \xrightarrow{\phi} \Gamma$$

$$\pi_1(M' - U_\eta) \longrightarrow \pi_1(M') \xrightarrow{\phi'} \Gamma$$

are identical and the composition

$$\pi_1(E_J) \longrightarrow \pi_1(M) \xrightarrow{\phi} \Gamma$$

factors through $H_1(E_J)$. Furthermore,

$$\rho(M, \phi) = \begin{cases} \rho(M', \phi'), & \text{if } \phi'(\eta) = 1, \\ \rho(M', \phi') + \rho(J), & \text{otherwise.} \end{cases}$$

The existence of ϕ and its uniqueness easily follow from the fact that $H_1(E_J)$ is an infinite cyclic group generated by the meridian of J. For the proof of the ρ-invariant formula, see [**14**, Proposition 3.2].

Returning to the construction of our example, choose a polynomial $\lambda(t)$ satisfying Lemma 5.20, and choose a ribbon knot K' with zero-surgery manifold of M' such that the rational Alexander module $H_1(M'; \mathbb{Q}[t^{\pm 1}])$ associated to $\phi'_0\colon \pi_1(M') \to \langle t \rangle$ sending the meridian of K' to t is isomorphic to $\mathbb{Q}[t^{\pm 1}]/\langle \lambda(t)^2 \rangle$. Choose a curve η in $S^3 - K'$ such that $\text{lk}(\eta, K') = 0$ and η represents the element

$$1 + \langle \lambda(t)^2 \rangle \in \mathbb{Q}[t^{\pm 1}]/\langle \lambda(t)^2 \rangle \cong H_1(M'; \mathbb{Q}[t^{\pm 1}])$$

We may assume that η is an unknotted simple closed curve by crossing change. Choose a knot J in S^3 such that the Arf invariant vanishes and $\rho(J) \neq 0$. For example, the connected sum of two trefoil knots has this property. By tying J along η as above, we obtain a new knot K.

5.4. KNOTS WHICH ARE NOT RATIONALLY (1.5)-SOLVABLE

THEOREM 5.23. *The knot K is integrally (1)-solvable but not rationally (1.5)-solvable.*

PROOF. Since $\eta \in [\pi_1(M'), \pi_1(M')]$ and J has vanishing Arf invariant, K is integrally (1)-solvable by Proposition 3.1 of [**14**].

Let M be the zero-surgery manifold of K and $\phi_0 \colon \pi_1(M) \to \langle t \rangle$ be the canonical map sending the meridian to t. Let denote the rational Alexander module $H_1(M; \mathbb{Q}[t^{\pm 1}])$ associated to ϕ_0 by $\mathcal{A}_0$. As in the above discussion, $M = (M' - U_\eta) \cup_\partial E_J$, and a standard Mayer–Vietoris argument shows that

$$M \longleftarrow (M' - U_\eta) \longrightarrow M'$$

gives rise to an isomorphism

$$\mathcal{A}_0 = H_1(M; \mathbb{Q}[t^{\pm 1}]) \cong H_1(M'; \mathbb{Q}[t^{\pm 1}]) = \mathbb{Q}[t^{\pm 1}]/\langle \lambda(t)^2 \rangle.$$

Let

$$\phi_1 = \phi_1(x_0, \phi_0) \colon \pi_1(M) \longrightarrow \Gamma_1,$$
$$\phi_1' = \phi_1'(x_0, \phi_0') \colon \pi_1(M') \longrightarrow \Gamma_1$$

be the homomorphisms determined by

$$x_0 = \lambda(t) + \langle \lambda(t)^2 \rangle \in \mathbb{Q}[t^{\pm 1}]/\langle \lambda(t)^2 \rangle.$$

Since K' is ribbon, $\rho(M', \phi_1') = 0$ by Proposition 5.21. We claim that $\rho(M, \phi_1)$ is nontrivial. Since η represents a generator of the cyclic module $\mathcal{A}_0$ and the Blanchfield form $B\ell_0$ on $\mathcal{A}_0$ is nondegenerated, $B\ell_0(\eta, x_0)$ is nontrivial. (Or alternatively, we can use the formula of the Blanchfield form given in Remark 5.19.) Therefore $\phi_1(\eta)$ is nontrivial (see the discussion on the induced homomorphism $\varphi = \varphi(x, \phi)$ at the beginning of 5.2.1). By the definition of our isomorphism between rational Alexander modules of K and K', we can apply Lemma 5.22 to obtain

$$\rho(M, \phi_1) = \rho(M', \phi_1') + \rho(J) = \rho(J) \neq 0.$$

From Proposition 5.21, it follows that K is not rationally (1.5)-solvable. □

By using different knots J, our construction produces infinitely many knots which are integrally (1)-solvable but not rationally (1.5)-solvable. In fact, if we use an appropriate family of knots J described below, it can be shown that these knots are linearly independent modulo rationally (1.5)-solvable knots.

PROPOSITION 5.24 (Proposition 2.6 in [**14**]). *There is a family of knots J_i, $i = 0, 1, 2 \ldots$, such that J_i has vanishing Arf invariants and the real numbers $\rho(J_i)$ are linearly independent over the integers.*

For a proof, see [**14**].

THEOREM 5.25. *There are infinitely many integrally (1)-solvable knots in S^3 which are linearly independent in $\mathcal{F}_{(1)}^{\mathbb{Q}}/\mathcal{F}_{(1.5)}^{\mathbb{Q}}$.*

PROOF. For $i = 1, 2, \ldots$, let K_i be the knot obtained by the construction of Theorem 5.23, using the knot J_i given in Proposition 5.24 in place of J. Since J_i has vanishing Arf invariant, K_i is integrally (1)-solvable as in the proof of Theorem 5.23.

We will show that the K_i are linearly independent in $\mathcal{F}^{\mathbb{Q}}_{(1)}/\mathcal{F}^{\mathbb{Q}}_{(1.5)}$. Suppose a linear combination

$$K = \#_{i=1}^{n} a_i K_i$$

is rationally (1.5)-solvable for some integers a_i, where $\#$ denotes the connected sum operation. We may assume that each a_i is positive by dropping vanishing terms and taking $-K_i$ instead of K_i if necessary. Since K_1 is not rationally (1.5)-solvable, we may also assume that $a_1 \geq 2$ if $n = 1$.

Let denote the zero surgery manifold of K_i and K by M_i and M, respectively. As in [14] and [28], we can construct a 4-manifold W with the following properties:

(1) $\partial W = M_1$ and W is a rational (1)-solution of K_1.
(2) For any homomorphism $\phi \colon \pi_1(M_1) \to \Gamma$ into a PTFA group Γ which extends to $\pi_1(W)$, the associated ρ-invariant is given by

$$\rho(M_1, \phi) = -\sum_{i=1}^{n} c_i \rho(J_i)$$

for some nonnegative integers c_i.

The arguments in [14, 28] construct an integral solution W with similar properties under an analogous integral solvability assumption. Since a minor modification of their argument constructs our W, we give a rough sketch only. By attaching 2-handles to the disjoint union $\bigcup a_i M_i \times [0, 1]$, we obtain a cobordism C from $\bigcup a_i M_i$ to M. Since each K_i is (1)-solvable, there exists an integral (1)-solution W_i of K_i. From the assumption that K is rationally (1.5)-solvable, there is a rational (1.5)-solution W_0 of K. Attaching W_0, C, $-(a_1 - 1)W_1$, $-a_2 W_2, \ldots, -a_n W_n$ along boundaries, we obtain a 4-manifold W with boundary M_1. In [28], it was shown that W is an integral (1)-solution when each W_0 is an integral solution. In our case, the same argument shows that W is a rational (1)-solution, whose complexity is the same as the that of W_0. Since ϕ factors through $\pi_1(W)$, $\rho(M_1, \phi)$ can be computed via the intersection form of W. As in [14, 28], since W_0 is a rational (1.5)-solution, W_0 has no contribution to the ρ-invariant by appealing to [13, Theorem 4.2]. So does C by a simple homological argument. So it suffices to consider the contribution from the W_i. Recall that K_i is obtained from a ribbon knot K' given in Theorem 5.18 by a satellite construction using J_i. Thus we have a specific (1)-solution W_i which is obtained by attaching a (0)-solution W'_i of J_i to the exterior W' of a slice disk of K' along a solid torus. As before, W' has no contribution to the ρ-invariant by appealing to [13, Theorem 4.2]. The contribution of W'_i is either trivial or

$\rho(J_i)$, depending to whether the image of the meridian of J_i is trivial in Γ. For more details, refer to [14, 28].

Now we use W to compute a particular ρ-invariant of M_1. Let c be the complexity of the rational solution W. Comparing with the integral case [14, 28], the main difficulty of our case is again that we do not know c. We control the metabolizer and the ρ-invariant as follows. Let

$$\phi_0, \phi_0' \colon \pi_1(M_1) \longrightarrow \Gamma_0 = \langle t \rangle$$

be the maps sending the meridian of K_1 to t, t^c, and let $\mathcal{A}_0$, $\mathcal{A}_0'$ be the associated rational Alexander modules, respectively. By the property (1) above and Theorem 5.13, there is a proper submodule P_0 in $\mathcal{A}_0'$ such that for any $x_0' \in P_0$, the associated map

$$\phi_1' = \phi_1'(x_0', \phi_0') \colon \pi_1(M_1) \longrightarrow \Gamma_1$$

factors through our rational solution W. On the other hand, as in the proof of Proposition 5.21, $\mathcal{A}_0 = \mathbb{Q}[t^{\pm 1}]/\langle \lambda(t)^2 \rangle$ and

$$\mathcal{A}_0' = \mathcal{A}_0 \otimes_{h_0} \mathcal{R}_0 = \mathbb{Q}[t^{\pm 1}]/\langle \lambda(t^c)^2 \rangle$$

has a unique proper submodule $\lambda(t)\mathcal{A}_0 \otimes_{h_0} \mathcal{R}_0$. Therefore P_0 must agree with this submodule, and in particular, we can think of $x_0' = x_0 \otimes 1$ where

$$x_0 = \lambda(t) + \langle \lambda(t)^2 \rangle \in \mathcal{A}_0.$$

Recall from the proof of Theorem 5.23 that $\rho(M_1, \phi_1) = \rho(J_1)$ where $\phi_1 = \phi_1(x_0, \phi_0)$. Combining this with the property (2) above and Corollary 5.17, we have

$$\rho(J_1) = \rho(M_1, \phi_1) = \rho(M_1, \phi_1') = -\sum_{i=1}^{n} c_i \rho(J_i).$$

This contradicts the linear independence of the $\rho(J_i)$. □

Bibliography

[1] R. C. Blanchfield, *Intersection theory of manifolds with operators with applications to knot theory*, Ann. of Math. (2) **65** (1957), 340–356.

[2] W. Browder, *Surgery on simply-connected manifolds*, Springer-Verlag, New York, 1972, Ergebnisse der Mathematik und ihrer Grenzgebiete, Band 65.

[3] S. E. Cappell and J. L. Shaneson, *The codimension two placement problem and homology equivalent manifolds*, Ann. of Math. (2) **99** (1974), 277–348.

[4] J. W. S. Cassels and A. Fröhlich (eds.), *Algebraic number theory*, Proceedings of an instructional conference organized by the London Mathematical Society (a NATO Advanced Study Institute) with the support of the International Mathematical Union, Academic Press, London, 1967.

[5] A. Casson and C. Gordon, *On slice knots in dimension three*, Algebraic and geometric topology (Proc. Sympos. Pure Math., Stanford Univ., Stanford, Calif., 1976), Part 2, Amer. Math. Soc., Providence, R.I., 1978, pp. 39–53.

[6] _____, *Cobordism of classical knots*, À la recherche de la topologie perdue, Birkhäuser Boston, Boston, MA, 1986, With an appendix by P. M. Gilmer, pp. 181–199.

[7] J. C. Cha and K. H. Ko, *Signatures of links in rational homology spheres*, Topology **41** (2002), no. 6, 1161–1182.

[8] _____, *Signature invariants of covering links*, Trans. Amer. Math. Soc. **358** (2006), no. 8, 3399–3412.

[9] J. Cheeger and M. Gromov, *Bounds on the von Neumann dimension of L^2-cohomology and the Gauss-Bonnet theorem for open manifolds*, J. Differential Geom. **21** (1985), no. 1, 1–34.

[10] T. D. Cochran and K. E. Orr, *Not all links are concordant to boundary links*, Bull. Amer. Math. Soc. (N.S.) **23** (1990), no. 1, 99–106.

[11] T. D. Cochran, *Noncommutative knot theory*, Algebr. Geom. Topol. **4** (2004), 347–398.

[12] T. D. Cochran and K. E. Orr, *Not all links are concordant to boundary links*, Ann. of Math. (2) **138** (1993), no. 3, 519–554.

[13] T. D. Cochran, K. E. Orr, and P. Teichner, *Knot concordance, Whitney towers and L^2-signatures*, Ann. of Math. (2) **157** (2003), no. 2, 433–519.

[14] _____, *Structure in the classical knot concordance group*, Comment. Math. Helv. **79** (2004), no. 1, 105–123.

[15] P. M. Cohn, *Free rings and their relations*, Academic Press, London, 1971, London Mathematical Society Monographs, No. 2.

[16] J. F. Davis and C. Livingston, *Alexander polynomials of periodic knots*, Topology **30** (1991), no. 4, 551–564.

[17] R. Fintushel and R. J. Stern, *A μ-invariant one homology 3-sphere that bounds an orientable rational ball*, Four-manifold theory (Durham, N.H., 1982), Contemp. Math., vol. 35, Amer. Math. Soc., Providence, RI, 1984, pp. 265–268.

[18] S. Friedl, *Eta invariants as sliceness obstructions and their relation to Casson-Gordon invariants*, Algebr. Geom. Topol. **4** (2004), 893–934 (electronic).

[19] S. L. Harvey, *Higher-order polynomial invariants of 3-manifolds giving lower bounds for the Thurston norm*, Topology **44** (2005), no. 5, 895–945.

[20] J. A. Hillman, *Alexander ideals of links*, Springer-Verlag, Berlin, 1981.

[21] M. W. Hirsch, *Embeddings and compressions of polyhedra and smooth manifolds*, Topology **4** (1966), 361–369.

[22] A. Kawauchi, *On links not cobordant to split links*, Topology **19** (1980), no. 4, 321–334.

[23] C. Kearton, *Blanchfield duality and simple knots*, Trans. Amer. Math. Soc. **202** (1975), 141–160.

[24] _____, *Cobordism of knots and Blanchfield duality*, J. London Math. Soc. (2) **10** (1975), no. 4, 406–408.

[25] _____, *The Milnor signatures of compound knots*, Proc. Amer. Math. Soc. **76** (1979), no. 1, 157–160.

[26] M. A. Kervaire, *Les nœuds de dimensions supérieures*, Bull. Soc. Math. France **93** (1965), 225–271.

[27] M. A. Kervaire and J. W. Milnor, *Groups of homotopy spheres. I*, Ann. of Math. (2) **77** (1963), 504–537.

[28] T. Kim, *Filtration of the classical knot concordance group and Casson-Gordon invariants*, Math. Proc. Cambridge Philos. Soc. **137** (2004), no. 2, 293–306.

[29] S. Lang, *Algebraic number theory*, second ed., Graduate Texts in Mathematics, vol. 110, Springer-Verlag, New York, 1994.

[30] J.-Y. Le Dimet, *Cobordisme d'enlacements de disques*, Mém. Soc. Math. France (N.S.) (1988), no. 32, ii+92.

[31] J. P. Levine, *Polynomial invariants of knots of codimension two*, Ann. of Math. (2) **84** (1966), 537–554.

[32] _____, *Invariants of knot cobordism*, Invent. Math. 8 (1969), 98–110; addendum, ibid. **8** (1969), 355.

[33] _____, *Knot cobordism groups in codimension two*, Comment. Math. Helv. **44** (1969), 229–244.

[34] R. A. Litherland, *Signatures of iterated torus knots*, Topology of low-dimensional manifolds (Proc. Second Sussex Conf., Chelwood Gate, 1977), Springer, Berlin, 1979, pp. 71–84.

[35] T. Matumoto, *On the signature invariants of a non-singular complex sesquilinear form*, J. Math. Soc. Japan **29** (1977), no. 1, 67–71.

[36] J. Milnor and D. Husemoller, *Symmetric bilinear forms*, Springer-Verlag, New York, 1973, Ergebnisse der Mathematik und ihrer Grenzgebiete, Band 73.

[37] J. W. Milnor, *Infinite cyclic coverings*, Conference on the Topology of Manifolds (Michigan State Univ., E. Lansing, Mich., 1967), Prindle, Weber & Schmidt, Boston, Mass., 1968, pp. 115–133.

[38] K. Murasugi, *On a certain numerical invariant of link types*, Trans. Amer. Math. Soc. **117** (1965), 387–422.

[39] F. Quinn, *Semifree group actions and surgery on PL homology manifolds*, Geometric topology (Proc. Conf., Park City, Utah, 1974), Springer, Berlin, 1975, pp. 395–414. Lecture Notes in Math., Vol. 438.

[40] A. Ranicki, *Exact sequences in the algebraic theory of surgery*, Mathematical Notes, vol. 26, Princeton University Press, Princeton, N.J., 1981.

[41] J. J. Rotman, *The theory of groups. An introduction*, second ed., Allyn and Bacon Inc., Boston, Mass., 1973, Allyn and Bacon Series in Advanced Mathematics.

[42] J.-P. Serre, *Local fields*, Graduate Texts in Mathematics, vol. 67, Springer-Verlag, New York, 1979, Translated from the French by Marvin Jay Greenberg.

[43] L. Taylor and B. Williams, *Local surgery: foundations and applications*, Algebraic topology, Aarhus 1978 (Proc. Sympos., Univ. Aarhus, Aarhus, 1978), Lecture Notes in Math., vol. 763, Springer, Berlin, 1979, pp. 673–695.

[44] A. G. Tristram, *Some cobordism invariants for links*, Proc. Cambridge Philos. Soc. **66** (1969), 251–264.

[45] H. F. Trotter, *Homology of group systems with applications to knot theory*, Ann. of Math. (2) **76** (1962), 464–498.

Editorial Information

To be published in the *Memoirs*, a paper must be correct, new, nontrivial, and significant. Further, it must be well written and of interest to a substantial number of mathematicians. Piecemeal results, such as an inconclusive step toward an unproved major theorem or a minor variation on a known result, are in general not acceptable for publication.

Papers appearing in *Memoirs* are generally at least 80 and not more than 200 published pages in length. Papers less than 80 or more than 200 published pages require the approval of the Managing Editor of the Transactions/Memoirs Editorial Board.

As of May 31, 2007, the backlog for this journal was approximately 15 volumes. This estimate is the result of dividing the number of manuscripts for this journal in the Providence office that have not yet gone to the printer on the above date by the average number of monographs per volume over the previous twelve months, reduced by the number of volumes published in four months (the time necessary for preparing a volume for the printer). (There are 6 volumes per year, each usually containing at least 4 numbers.)

A Consent to Publish and Copyright Agreement is required before a paper will be published in the *Memoirs*. After a paper is accepted for publication, the Providence office will send a Consent to Publish and Copyright Agreement to all authors of the paper. By submitting a paper to the *Memoirs*, authors certify that the results have not been submitted to nor are they under consideration for publication by another journal, conference proceedings, or similar publication.

Information for Authors

Memoirs are printed from camera copy fully prepared by the author. This means that the finished book will look exactly like the copy submitted.

Initial submission. The AMS uses Centralized Manuscript Processing for initial submissions. Authors should submit a PDF file using the Initial Manuscript Submission form found at www.ams.org/cgi-bin/peertrack/submission.pl, or send one copy of the manuscript to the following address: Centralized Manuscript Processing, MEMOIRS OF THE AMS, 201 Charles Street, Providence, RI 02904-2294 USA. If a paper copy is being forwarded to the AMS, indicate that it is for it Memoirs and include the name of the corresponding author, contact information such as email address or mailing address, and the name of an appropriate Editor to review the paper (see the list of Editors below).

The paper must contain a *descriptive title* and an *abstract* that summarizes the article in language suitable for workers in the general field (algebra, analysis, etc.). The *descriptive title* should be short, but informative; useless or vague phrases such as "some remarks about" or "concerning" should be avoided. The *abstract* should be at least one complete sentence, and at most 300 words. Included with the footnotes to the paper should be the 2000 *Mathematics Subject Classification* representing the primary and secondary subjects of the article. The classifications are accessible from www.ams.org/msc/. The list of classifications is also available in print starting with the 1999 annual index of *Mathematical Reviews*. The Mathematics Subject Classification footnote may be followed by a list of *key words and phrases* describing the subject matter of the article and taken from it. Journal abbreviations used in bibliographies are listed in the latest *Mathematical Reviews* annual index. The series abbreviations are also accessible from www.ams.org/publications/. To help in preparing and verifying references, the AMS offers MR Lookup, a Reference Tool for Linking, at www.ams.org/mrlookup/.

Electronically prepared manuscripts. The AMS encourages electronically prepared manuscripts, with a strong preference for $\mathcal{AMS}$-LaTeX. To this end, the Society has prepared $\mathcal{AMS}$-LaTeX author packages for each AMS publication. Author packages include instructions for preparing electronic manuscripts, samples, and a style file that generates the particular design specifications of that publication series. Though $\mathcal{AMS}$-LaTeX is the highly preferred format of TeX, author packages are also available in $\mathcal{AMS}$-TeX.

Authors may retrieve an author package from the AMS website starting from www.ams.org/tex/ or via FTP to ftp.ams.org (login as anonymous, enter username as password, and type cd pub/author-info). The *AMS Author Handbook* and the *Instruction Manual* are available in PDF format following the author packages link from www.ams.org/tex/. The author package can also be obtained free of charge by sending email to tech-support@ams.org (Internet) or from the Publication Division, American Mathematical Society, 201 Charles St., Providence, RI 02904-2294, USA. When requesting an author package, please specify $\mathcal{AMS}$-LaTeX or $\mathcal{AMS}$-TeX and the publication in which your paper will appear. Please be sure to include your complete mailing address.

After acceptance. The final version of the electronic file should be sent to the Providence office (this includes any TeX source file, any graphics files, and the DVI or PostScript file) immediately after the paper has been accepted for publication.

Before sending the source file, be sure you have proofread your paper carefully. The files you send must be the EXACT files used to generate the proof copy that was accepted for publication. For all publications, authors are required to send a printed copy of their paper, which exactly matches the copy approved for publication, along with any graphics that will appear in the paper.

Accepted electronically prepared files can be submitted via the web at www.ams.org/submit-book-journal/, sent via FTP, or sent on CD-Rom or diskette to the Electronic Prepress Department, American Mathematical Society, 201 Charles Street, Providence, RI 02904-2294 USA. TeX source files, DVI files, and PostScript files can be transferred over the Internet by FTP to the Internet node ftp.ams.org (130.44.1.100). When sending a manuscript electronically via CD-Rom or diskette, please be sure to include a message identifying the paper as a Memoir.

Electronically prepared manuscripts can also be sent via email to pub-submit@ams.org (Internet). In order to send files via email, they must be encoded properly. (DVI files are binary and PostScript files tend to be very large.)

Electronic graphics. Comprehensive instructions on preparing graphics are available at www.ams.org/jourhtml/. A few of the major requirements are given here.

Submit files for graphics as EPS (Encapsulated PostScript) files. This includes graphics originated via a graphics application as well as scanned photographs or other computer-generated images. If this is not possible, TIFF files are acceptable as long as they can be opened in Adobe Photoshop or Illustrator. No matter what method was used to produce the graphic, it is necessary to provide a paper copy to the AMS.

Authors using graphics packages for the creation of electronic art should also avoid the use of any lines thinner than 0.5 points in width. Many graphics packages allow the user to specify a "hairline" for a very thin line. Hairlines often look acceptable when proofed on a typical laser printer. However, when produced on a high-resolution laser imagesetter, hairlines become nearly invisible and will be lost entirely in the final printing process.

Screens should be set to values between 15% and 85%. Screens which fall outside of this range are too light or too dark to print correctly. Variations of screens within a graphic should be no less than 10%.

Inquiries. Any inquiries concerning a paper that has been accepted for publication should be sent to memo-query@ams.org or directly to the Electronic Prepress Department, American Mathematical Society, 201 Charles St., Providence, RI 02904-2294 USA.

Editors

This journal is designed particularly for long research papers, normally at least 80 pages in length, and groups of cognate papers in pure and applied mathematics. Papers intended for publication in the *Memoirs* should be addressed to one of the following editors. The AMS uses Centralized Manuscript Processing for initial submissions to AMS journals. Authors should follow instructions listed on the Initial Submission page found at www.ams.org/memo/memosubmit.html.

Algebra to ALEXANDER KLESHCHEV, Department of Mathematics, University of Oregon, Eugene, OR 97403-1222; email: ams@noether.uoregon.edu

Algebraic geometry and its application to MINA TEICHER, Emmy Noether Research Institute for Mathematics, Bar-Ilan University, Ramat-Gan 52900, Israel; email: teicher@macs.biu.ac.il

Algebraic geometry to DAN ABRAMOVICH, Department of Mathematics, Brown University, Box 1917, Providence, RI 02912; email: amsedit@math.brown.edu

Algebraic number theory to V. KUMAR MURTY, Department of Mathematics, University of Toronto, 100 St. George Street, Toronto, ON M5S 1A1, Canada; email: murty@math.toronto.edu

Algebraic topology to ALEJANDRO ADEM, Department of Mathematics, University of British Columbia, Room 121, 1984 Mathematics Road, Vancouver, British Columbia, Canada V6T 1Z2; email: adem@math.ubc.ca

Combinatorics to JOHN R. STEMBRIDGE, Department of Mathematics, University of Michigan, Ann Arbor, Michigan 48109-1109; email: FRS@umich.edu

Complex analysis and harmonic analysis to ALEXANDER NAGEL, Department of Mathematics, University of Wisconsin, 480 Lincoln Drive, Madison, WI 53706-1313; email: nagel@math.wisc.edu

Differential geometry and global analysis to LISA C. JEFFREY, Department of Mathematics, University of Toronto, 100 St. George St., Toronto, ON Canada M5S 3G3; email: jeffrey@math.toronto.edu

Dynamical systems and ergodic theory to AMIE WILKINSON, Department of Mathematics, Northwestern University, 2033 Sheridan Road, Evanston, IL 60208-2730; email: transactions@math.northwestern.edu

Functional analysis and operator algebras to DIMITRI SHLYAKHTENKO, Department of Mathematics, University of California, Los Angeles, CA 90095; email: shlyakht@math.ucla.edu

Geometric analysis to WILLIAM P. MINICOZZI II, Department of Mathematics, Johns Hopkins University, 3400 N. Charles St., Baltimore, MD 21218; email: trans@math.jhu.edu

Geometric analysis to MLADEN BESTVINA, Department of Mathematics, University of Utah, 155 South 1400 East, JWB 233, Salt Lake City, Utah 84112-0090; email: bestvina@math.utah.edu

Harmonic analysis, representation theory, and Lie theory to ROBERT J. STANTON, Department of Mathematics, The Ohio State University, 231 West 18th Avenue, Columbus, OH 43210-1174; email: stanton@math.ohio-state.edu

Logic to STEFFEN LEMPP, Department of Mathematics, University of Wisconsin, 480 Lincoln Drive, Madison, Wisconsin 53706-1388; email: lempp@math.wisc.edu

Partial differential equations to GUSTAVO PONCE, Department of Mathematics, South Hall, Room 6607, University of California, Santa Barbara, CA 93106; email: ponce@math.ucsb.edu

Partial differential equations and dynamical systems to PETER POLACIK, School of Mathematics, University of Minnesota, Minneapolis, MN 55455; email: polacik@math.umn.edu

Probability and statistics to KRZYSZTOF BURDZY, Department of Mathematics, University of Washington, Box 354350, Seattle, Washington 98195-4350; email: burdzy@math.washington.edu

Real analysis and partial differential equations to DANIEL TATARU, Department of Mathematics, University of California, Berkeley, Berkeley, CA 94720; email: tataru@math.berkeley.edu

All other communications to the editors should be addressed to the Managing Editor, ROBERT GURALNICK, Department of Mathematics, University of Southern California, Los Angeles, CA 90089-1113; email: guralnic@math.usc.edu.

Titles in This Series

887 **Charlotte Wahl,** Noncommutative Maslov index and eta-forms, 2007

886 **Robert M. Guralnick and John Shareshian,** Symmetric and alternating groups as monodromy groups of Riemann surfaces I: Generic covers and covers with many branch points, 2007

885 **Jae Choon Cha,** The structure of the rational concordance group of knots, 2007

884 **Dan Haran, Moshe Jarden, and Florian Pop,** Projective group structures as absolute Galois structures with block approximation, 2007

883 **Apostolos Beligiannis and Idun Reiten,** Homological and homotopical aspects of torsion theories, 2007

882 **Lars Inge Hedberg and Yuri Netrusov,** An axiomatic approach to function spaces, spec tral synthesis and Luzin approximation, 2007

881 **Tao Mei,** Operator valued Hardy spaces, 2007

880 **Bruce C. Berndt, Geumlan Choi, Youn-Seo Choi, Heekyoung Hahn, Boon Pin Yeap, Ae Ja Yee, Hamza Yesilyurt, and Jinhee Yi,** Ramanujan's forty identities for Rogers-Ramanujan functions, 2007

879 **O. García-Prada, P. B. Gothen, and V. Muñoz,** Betti numbers of the moduli space of rank 3 parabolic Higgs bundles, 2007

878 **Alessandra Celletti and Luigi Chierchia,** KAM stability and celestial mechanics, 2007

877 **María J. Carro, José A. Raposo, and Javier Soria,** Recent developments in the theory of Lorentz spaces and weighted inequalities, 2007

876 **Gabriel Debs and Jean Saint Raymond,** Borel liftings of Borel sets: Some decidable and undecidable statements, 2007

875 **C. Krattenthaler and T. Rivoal,** Hypergéométrie et fonction zêta de Riemann, 2007

874 **Sonia Natale,** Semisolvability of semisimple Hopf algebras of low dimension, 2007

873 **A. J. Duncan,** Exponential genus problems in one-relator products of groups, 2007

872 **Anthony V. Geramita, Tadahito Harima, Juan C. Migliore, and Yong Su Shin,** The Hilbert function of a level algebra, 2007

871 **Pascal Auscher,** On necessary and sufficient conditions for L^p-estimates of Riesz transforms associated to elliptic operators on $\mathbb{R}^n$ and related estimates, 2007

870 **Takuro Mochizuki,** Asymptotic behaviour of tame harmonic bundles and an application to pure twistor D-modules, Part 2, 2007

869 **Takuro Mochizuki,** Asymptotic behaviour of tame harmonic bundles and an application to pure twistor D-modules, Part 1, 2007

868 **Gelu Popescu,** Entropy and multivariable interpolation, 2006

867 **Vilmos Totik,** Metric properties of harmonic measures, 2006

866 **William Craig,** Semigroups underlying first-order logic, 2006

865 **Nathanial P. Brown,** Invariant means and finite representation theory of $C*$-algebras, 2006

864 **John M. Lee,** Fredholm operators and Einstein metrics on conformally compact manifolds, 2006

863 **M. Lübke and A. Teleman,** The Universal Kobayashi-Hitchin correspondence on Hermitian manifolds, 2006

862 **Alberto Canonaco,** The Beilinson complex and canonical rings of irregular surfaces, 2006

861 **Leon A. Takhtajan and Lee-Peng Teo,** Weil-Petersson metric on the universal Teichmüller space, 2006

860 **Thomas M. Fiore,** Pseudo limits, biadjoints and pseudo algebras: Categorical foundations of conformal field theory, 2006

859 **N. Arcozzi, R. Rochberg, and E. Sawyer,** Carleson measures and interpolating sequences for Besov spaces on complex balls, 2006

TITLES IN THIS SERIES

858 Enrico Valdinoci, Berardino Sciunzi, and Vasile Ovidiu Savin, Flat level set regularity of p-Laplace phase transitions, 2006

857 Donatella Danielli, Nocola Garofalo, and Duy-Minh Nhieu, Non-doubling Ahlfors measures, perimeter measures, and the characterization of the trace spaces of Sobolev functions in Carnot-Carathéodory spaces, 2006

856 Vladimir Bolotnikov and Harry Dym, On boundary interpolation for matrix valued Schur functions, 2006

855 Yevgenia Kashina, Yorck Sommerhäuser, and Yongchang Zhu, On higher Frobenius-Schur indicators, 2006

854 Noam Greenberg, The role of true finiteness in the admissible recursively enumerable degrees, 2006

853 Joachim Krieger, Stability of spherically symmetric wave maps, 2006

852 Viorel Barbu, Irena Lasiecka, and Roberto Triggiani, Tangential boundary stabilization of Navier-Stokes equations, 2006

851 Jie Wu, On maps from loop suspensions to loop spaces and the shuffle relations on the Cohen groups, 2006

850 Siegfried Echterhoff, S. Kaliszewski, John Quigg, and Iain Raeburn, A categorical approach to imprimitivity theorems for C^*-dynamical systems, 2006

849 Katsuhiko Kuribayashi, Mamoru Mimura, and Tetsu Nishimoto, Twisted tensor products related to the cohomology of the classifying spaces of loop groups, 2006

848 Bob Oliver, Equivalences of classifying spaces completed at the prime two, 2006

847 Eric T. Sawyer and Richard L. Wheeden, Hölder continuity of weak solutions to subelliptic equations with rough coefficients, 2006

846 Victor Beresnevich, Detta Dickinson, and Sanju Velani, Measure theoretic laws for lim–sup sets, 2006

845 Ehud Friedgut, Vojtech Rödl, Andrzej Ruciński, and Prasad V. Tetali, A Sharp threshold for random graphs with a monochromatic triangle in every edge coloring, 2006

844 Amadeu Delshams, Rafael de la Llave, and Tere M. Seara, A geometric mechanism for diffusion in Hamiltonian systems overcoming the large gap problem: Heuristics and rigorous verification on a model, 2006

843 Denis V. Osin, Relatively hyperbolic groups: Intrinsic geometry, algebraic properties, and algorithmic problems, 2006

842 David P. Blecher and Vrej Zarikian, The calculus of one-sided M-ideals and multipliers in operator spaces, 2006

841 Enrique Artal Bartolo, Pierrette Cassou-Noguès, Ignacio Luengo, and Alejandro Melle Hernández, Quasi-ordinary power series and their zeta functions, 2005

840 Sławomir Kołodziej, The complex Monge-Ampère equation and pluripotential theory, 2005

839 Mihai Ciucu, A random tiling model for two dimensional electrostatics, 2005

838 V. Jurdjevic, Integrable Hamiltonian systems on complex Lie groups, 2005

837 Joseph A. Ball and Victor Vinnikov, Lax-Phillips scattering and conservative linear systems: A Cuntz-algebra multidimensional setting, 2005

836 H. G. Dales and A. T.-M. Lau, The second duals of Beurling algebras, 2005

835 Kiyoshi Igusa, Higher complex torsion and the framing principle, 2005

834 Kenîchi Ohshika, Kleinian groups which are limits of geometrically finite groups, 2005

For a complete list of titles in this series, visit the AMS Bookstore at **www.ams.org/bookstore/**.